Wireless in Wonderland

Canadians Cut The Cord

CANTEL
1983–1993

George A. Fierheller

One Approach to a Mobile Phone

Library and Archives Canada Cataloguing in Publication

Fierheller, George A., 1933-
Wireless in wonderland : Canadians cut the cord : CANTEL 1983-1993 /
George A. Fierheller.

Includes bibliographical references.
ISBN 978-1-897486-11-5

1. Cellular telephone services industry--Canada.
2. Wireless communication systems--Canada.
3. Cantel (Firm). 4. Fierheller, George A., 1933-. I. Title.

HE9715.C2F53 2010 384.5'350971 C2010-900332-2

Cover design by Lina Di Nardo Moss
tdmp@rogers.com

Produced and printed by
Stewart Publishing & Printing
Markham, Ontario, Canada L3P 2X3
Tel: 905-294-4389
robert@stewartbooks.com www.stewartbooks.com

"History will treat me kindly
for I intend to write it"

- Winston Churchill

"Me too"

- George Fierheller

"I've always been a perfectionist – just not a very good one."

It is easy if you look at it this way...

What's Inside

PART TWO
Let The Games Begin

PART THREE
Back In The Cellular Saddle

Dedication

Edward S. (Ted) Rogers

As has often been said, success has many claimants, but no one is more directly responsible for the ultimate success of what was then called CANTEL, now Rogers Wireless, than the late Ted Rogers.

Where others' faith faltered from time to time, Ted remained dedicated to the new wireless technology and the company that he and the other founders started. It is now the largest wireless communications organization in Canada, with more than eight million subscribers.

Ted was the one most responsible for getting me involved in this incredible adventure. He and I had known each other since our college days at Trinity, Sigma Chi and elsewhere.

He is remembered by everyone as one of Canada's leading entrepreneurs – and by me as a wonderful friend.

Acknowledgements

As with any of my books, I take full responsibility for the ultimate text. There is no editor to blame and all of the opinionated opinions are mine.

However, a number of people were very helpful in providing material from their own files. In particular, I would mention Roger Keay, recently retired Vice President, Technology at Rogers. Roger was one of the first employees of CANTEL. Also Bob Berner, Executive Vice President and Chief Technology Officer, Rogers Wireless, was one of CANTEL's pioneers, and has been a good sounding board.

Jeannie Hastie, the long-time Assistant to Ted Rogers, also was most helpful.

Jeannie Hastie with Ted

In 1994, Alistair Dow was asked by Rogers Communications to pull together some notes on the early history of CANTEL. He did this although the company never published any of it. He was kind enough to send me copies and I have used parts of them from time to time.

Finally, I would like to thank Connie Nesbitt who patiently typed my dictated text, and Robert Stewart, my publisher, who turned the text into a readable document.

OTHER STUFF

Some of my earlier books may be of interest. *Finnie's Family* and *Let Me Say This About That* contain references to CANTEL. All books shown on these two pages are available in printable form at www.gfierheller.ca

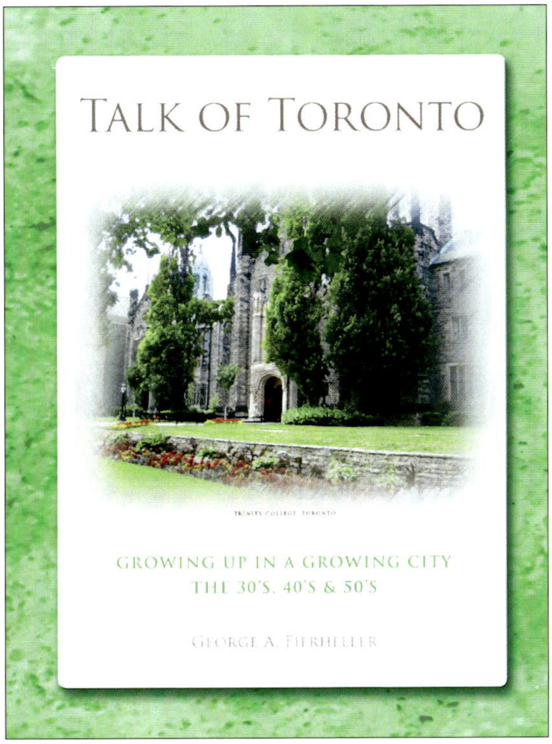

The SDL STORY

The First 125 Years: The National Club

The Toronto Adventurer's Club - 25th Anniversary

Cantel in Context

Corporate histories can be nothing more than recycled archives. Even when the story is about the remarkable rise of a company in a remarkable industry, it runs the risk of being a pretty dull recitation of the facts. There is no guarantee I will do any better, but I will concentrate on the personalities involved and the way their interaction helped the company get where it is and, on occasion, nearly put it under.

With the passing of Ted Rogers in December, 2008 it seems timely to record some recollections.

There are a couple of excellent books that do cover some of the facts about the development of CANTEL and its evolution into Rogers Wireless.

Caroline Van Hasselt in her book, *High Wire Act: Ted Rogers And The Empire That Debt Built*, contains a wonderfully detailed story of the birth of CANTEL and I highly recommend it.

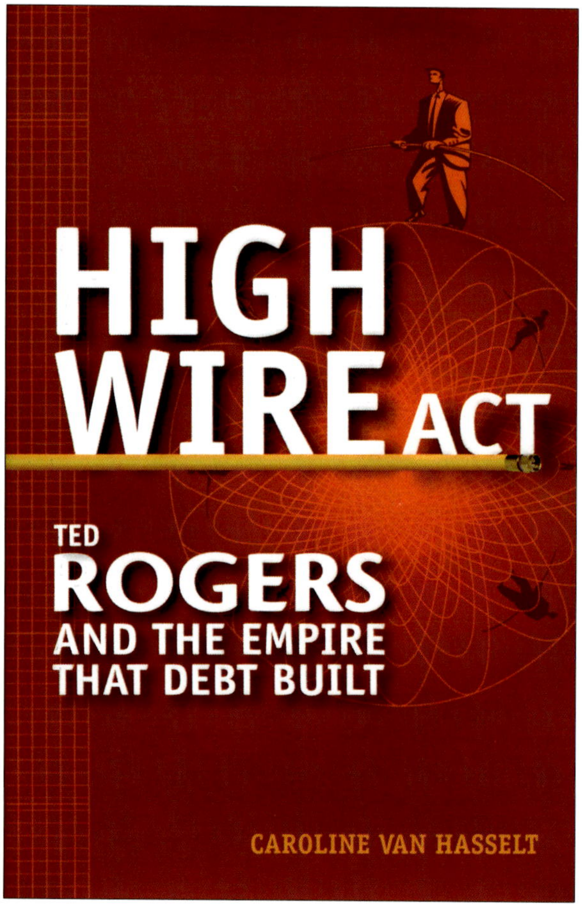

High Wire Act and Relentless

Ted's own autobiography, *Relentless*, provides an even more personal view of this and the many other successes in his life. It is a wonderful read.

I will try to supplement rather than duplicate those books. But, surprisingly, there appears to be very little else written about the transition from the Wired World to the Wireless World in Canada. Obviously, CANTEL was at the very centre of all of this as Canada's first national telephone company.

I was not the entrepreneur in this story, although, as we shall see, I was involved right from the start – both leading the team to get the licences for the non-wireline cellular frequencies and was then CANTEL's founding President, Chairman and/or CEO in various combinations for the first 10 years of its existence. No wonder I have very little hair left!

A Peek At People

Corporations have their own personality and CANTEL was no different. However, the personality of a company is largely the sum of the personalities of those involved – particularly in its early days.

CANTEL Annual Report 1993

CANTEL was lucky enough to rise above its schizophrenic start. The personalities involved had all the characteristics of real entrepreneurs – they were strong-minded, bright, egotistical, determined and largely ungovernable.

The company succeeded because of this and in spite of this. When you think that CANTEL was launched from a background almost devoid of telephone experience and in competition with some of the largest entrenched firms in Canada, it is a remarkable success story.

Today Rogers Wireless is the largest and most successful mobile communications company in the country. It is owned by one organization, Rogers Communications Inc.

Its main competitor in the initial stages was Bell Canada, which had been in existence for more than 100 years and was, at the time, Canada's largest and most profitable corporation. Of course CANTEL also had to compete with the local telephone company in each jurisdiction, e.g., BC Tel in British Columbia, AGT in Alberta and so on, but, looking just at Bell, which Ted referred to as the 'evil empire', the more remarkable thing is that Rogers Communications Inc. has survived and ultimately thrived, while Bell has lurched from problem to problem.

This reversal of fortune took place in a little over 20 years. And yet Bell has had the leadership of many of the brightest minds in Canada, including the likes of Jean de Grandpré, Ray Cyr, Jean Monty, John McLennan (of whom we will read more), Red Wilson and Michael Sabia to mention a few. They just don't come any better.

So what happened? I will try to answer that, amongst other things, from the point of view of my involvement in CANTEL's initial decade from 1983 to 1993.

But before I do, it is important to look at the overall development of the communications field to put this story in context.

Here goes.

1947 Trans Canada Telephone System

PART ONE

In The Beginning

STARTING AT THE START

My wife, Glenna, and I were on a tour with the Chief Executives Organization visiting the Baltic countries. In Latvia, CEO had arranged for an organ concert in an old church. Glenna thinks that organs are out-of-tune bagpipes but, despite being an atheist, I like organ music.

I was seated in one of the pews with Glenna on my left and a quite attractive 40-somethingish widow on my right. CEO encourages widows or widowers of CEO members to continue to join the group on trips.

She was what I would call a 'gusher'.

The music had no sooner started than she also started.

"Oh, the church is so beautiful."

"Yeah, beautiful," I replied.

"Oh, the windows are so ethereal."

"Yeah, ethereal," I replied, again trying to put an end to the conversation.

Baltic View

"Oh, I just love old organs."
"That's nice because I have one of those."
Glenna was trying to shut me up, but the gusher continued.
"Oh, do you?"
"Lady, don't go there."
Glenna by this time was elbowing me to knock it off.

Wireless communication. Or senseless communication. Either way it illustrates my point. For as long as human beings have communicated, it has been through means other than wires. It was Samuel Morse and Alexander Graham Bell who convinced us there was a better way. It was, however, only in the 1830s that Morse developed the means of communicating over a wire, using the Morse Code that he introduced in 1838.

Prior to this, communicating at a distance, whether a few metres or even kilometres, had to be done with some form of wireless communication.

Our Darwinian ancestors likely communicated by making gestures even before they developed rudimentary oral communications (I wonder who developed the one-finger salute – this certainly preceded Pierre Elliott Trudeau).

There is some dispute about whether the Neanderthals had the ability to speak as we understand the term, but there are some indications that *Homo erectus* had developed some primitive ability at oral communication perhaps one million years ago.

This is really starting at the start, but it is to make the point that wireless communication is normal, and it is communication over a wire that is out of step.

While we are reminiscing about the good old days, it is important to remember that there were many other ways of communicating than just speech. One only has to look at the unbelievable cave paintings, such as those at Chauvet that date back perhaps

The Panel of Horses, Chauvet Cave

30,000 years, to understand the desire of our ancestors to communicate as best they could.

The sensitivity exhibited in these early cave paintings was developed even further in the millennia that followed. The recent Royal Ontario Museum exhibition of Trypilian Culture shows remarkably advanced communication through not only painting and decoration but also with sculpting. The wistful statue of a woman, dating possibly from 7,000 years ago, communicates with us today with just a glance. What is she trying to communicate? Although clearer communication would have to wait for the development of writing, this ancient small figurine clearly communicates emotions across the centuries.

Roger Fouts, who has worked extensively on the language capability of higher primates, notes that "80% of the meaning of a face-to-face conversation passes nonverbally." Is this the reason for the huge popularity of digital images on contemporary cell phones? We should have seen that one coming but I, for one, did not.

Writing was another early form of communication. The Sumerians took the lead with the extensive use of pictograms, and by around 3700 BC these pictograms started to evolve into a more abstract system made with a wedge-shaped tool pressed into the damp clay. By 2500 BC cuneiform had settled in to a system of writing in which each sign represented composite sounds. Oral communication was now linked to the written record.

This compact form of communication, which has existed for almost 5,000 years, might have also given us a clue to the human desire for text messaging on wireless. Again, this was not foreseen in the early wireless systems but probably should have been, given the long history of this kind of communication. Blackberry here we come.

The whole history of pre-Morse wireless communication is fascinating, but well beyond the scope of this book. Before we leave the area of history with all its precedents for later wireless technology, it is worth touching on a few other interesting precedents.

Trypilian Figurine

Joshua was certainly communicating with the inhabitants of Jericho when he blew on what was likely a ram's horn, allegedly knocking down the wall. In all likelihood this simply meant he was summoning his troops. But this was a form of communication, albeit analogue.

It was not long, however, before digital communication was developed. The drums used in Africa effectively sent their signals in digital form.

There are stories from the Greek and Roman era of polished shields being used to reflect the sunlight to transmit what were really digital messages over long distances.

North American Indians developed reasonably complex smoke signals with the same purpose. They were not alone in using this technique.

Having mentioned a ram's horn, we should also bear in mind that music was and is a very universal way of communicating. It is not clear just when music began to play a role in human society, but certainly by the time of the Greeks it was well established. In fact, according to Burkholder, Grout and Palisca in the *History of Western Music*, the Greeks actually believed that music could affect one's ethical character or way of behaving. No one today would doubt they were correct in this.

English Medieval Music - Sumer is icuman in

It is also interesting to remember that the notation used to put music on paper has evolved into the only true international language, recognizable in virtually all communities.

Again, we should not have been surprised to see the modern wireless world quickly adapt to everything – from ring tones to downloading music to iPods.

Gestures, of course, do tend to have international recognition as a means of communication. A raised hand with the palm facing forward is almost universally recognized to mean 'stop'. A clenched fist could hardly be misinterpreted.

But many forms of communication were not meant to be universally recognized. In fact, nations went to great lengths to encrypt information to limit who could make sense of a message.

However, the desire to transmit messages over long distances, whether encrypted or not, led to a number of ingenious solutions.

In 1792, a French engineer, Claude Chappe, built the first visual semaphore system between Lille and Paris. It involved an ingenious set of pulleys rotating beams of wood on the top of a tower, the positions of which would indicate the information. A series of towers in line of sight had to be built so the message could be relayed from tower to tower. It was essentially the same principle as the manual semaphore flags used to transmit alphabetical messages over short line-of-sight distances, primarily for military use.

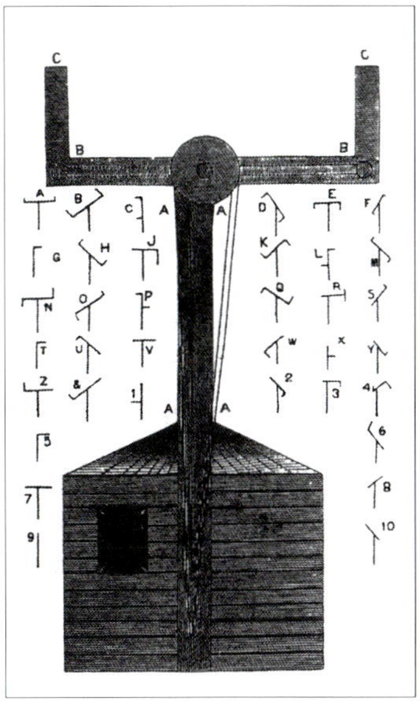

Schematic of a Chappe Optical Telegraph

A replica of a Chappe Tower

The system was obviously expensive as towers were required every 10 to 30 kilometres, and very skilled operators were required to manipulate the signal pulleys. As a result, the last commercial line was abandoned in 1880.

However, there was another reason why these early attempts to transmit information over long distances were abandoned. On January 6, 1838, Samuel Morse successfully demonstrated his electrical telegraph over a distance of 5 kilometres.

Samuel F.B. Morse

Soon after he was joined by Alfred Vail, who developed a means of logging the newly developed Morse Code on paper tape.

Long-distance electrical communication was on its way, albeit over wires for the first time.

And Science proclaimed, from shore to shore,
That Time and Space ruled man no more.
– The Victory
(a tribute to Samuel Morse, 1872)
The Victorian Internet
– Tom Standage, 1992

The Great Eastern at Heart's Content Harbour, Newfoundland, after laying the first transatlantic cable

A Wired World

By 1844, Morse had demonstrated his system successfully between Washington DC and Baltimore. By 1851 telegraph lines in the United States spanned some 32,000 kilometres.

Samuel F.B. Morse was one of many colourful characters in the early history of communications. He began life as a painter and was a founder of the National Academy of Design and its first President from 1826 to 1845. While the idea of an electric telegraph built around the newly discovered electromagnet was hardly new (having been put forward even before 1800) and while similar parallel efforts were going on in England, it was Morse who designed the code that made the system practical. He also turned out to be a promoter who knew how to get recognition for his device. Even the first message sent in 1844 over the Baltimore/Washington circuit had the element of hucksterism. It said, "What hath God wrought!"

A typical Key designed by J. D. Caton

What Morse did not appreciate, however, was that he was fundamentally changing the way people communicate. Wires went from place to place, not from person to person. But more was to come.

While land lines to carry the telegraphic signals were being strung everywhere, the first transatlantic telegraph cable was completed on July 27, 1866, allowing useful transatlantic telecommunication for the first time. There actually had been earlier transatlantic cables installed in 1857 and 1858, but they only operated for a few days before failing.

In the realm of pure trivia, the first successful transatlantic cable was laid by *The Great Eastern*. This was a remarkable steamship and, at the time of its launching in 1858, it was the largest ship in the world, displacing nearly 19,000 tonnes. It actually was a hybrid – with two paddle engines, two screw engines and six masts of sails. However, the huge ship was never a commercial success, and in 1864 it was converted to a cable-laying vessel. This prototype of the modern ocean liner was broken up in 1889.

But if the telegraph started our love affair with the wired world, it was the invention of the telephone that popularized wireline transmission. Few could operate a telegraph key. Everyone could speak however, and they did so at length.

For Whom The Bell Tolls

The story of Alexander Graham Bell and the invention of the telephone is too well known to require much elaboration. Typically, when an invention has the impact of the telephone, many countries claim its inventor as their own. Bell was born in Edinburgh, the son and grandson of a remarkable family of speech specialists. The young Bell had been educated in London as well as Edinburgh, and there his interest in the possible transmission of sound over a wire was whetted by a chance meeting with Sir Charles

Wheatstone, the best-known English expert on telegraphy. In fact, Sir Charles constructed the first commercial electrical telegraph in England only a few months after Samuel Morse's successful demonstration of such a device on January 6, 1838, and Sir Charles started his commercial operation on April 9, 1839. Bell was only 22 when he met Sir Charles, who was by then 67 and world famous but the young Bell never forgot the encounter.

Journal of the Telegraph, June 16, 1873

As is well known to all Canadians, for some time he made his home in Brantford, Ontario, although the Americans have staked their claim to being the country of origin of the telephone because of the experiment held in Boston in June, 1875, where Bell and Thomas A. Watson, his assistant, distinctly heard the twang of a clock spring that had been transmitted over the wire. It had not yet spoken a word but Bell, who had spent his early career working with deaf-mutes (including his future wife Mabel Hubbard), had allegedly said, "if I can make a deaf-mute talk, I can make iron talk."

It was on March 10, 1876, after considerably more tinkering with the invention, that the phrase that was to be heard around the world was first heard by Watson, "Mr. Watson, come here I want you." Watson allegedly ran up three flights of stairs shouting, "I can hear you. I can hear the words."

It was on his 29th birthday that Bell received Patent #174,465 – the most valuable single patent ever issued in any country.

Bell's Telephone, 1876

For all this the telephone got very little early attention

until it was demonstrated at the Centennial Exposition in Philadelphia two months later. Bell almost did not go, as he was still an impoverished inventor.

The story is told in Herbert N. Casson's *The History of the Telephone* about how the Emperor of Brazil, Dom Pedro de Alcantara and his wife, the Empress Theresa, walked into the exhibition room at the same time that some of the judges, evaluating various new products at the Exposition, were there. Dom Pedro exclaimed, "Professor Bell, I am delighted to see you again." The judges had no idea how Bell would know anyone so famous, but it turns out that Dom Pedro had once visited one of Bell's classes for deaf-mutes at Boston University and, being a humanitarian, Bell had recently helped to organize the first Brazilian school for deaf-mutes in Rio de Janeiro.

Dom Pedro picked up the receiver, listened for a moment and exclaimed, "My God – it talks!"

Casson's The History of the Telephone, 1910

News of the incident spread quickly. Professor Joseph Henry, who had encouraged Bell to keep on with his experiments, was the dean of the theory of electrical science in the United States and listened to the device with awe.

He was followed by Sir William Thomson from the United Kingdom. Sir William was probably the foremost electrical scientist in the world at the time and had been the engineer of the first Atlantic cable. He is best remembered as Lord Kelvin. His comment was, "It does speak, it is the most wonderful thing I have seen in America."

There was obviously no looking back.

By the time Casson wrote his history (published June 1, 1910) 35 years had passed since the first telephone call. At that time there were already more than 10 million telephones scattered around the world, with seven million in the United States alone. All were connected by wires.

The Wired World

The wires, whether for telegraph or the new telephone, were everywhere. The next challenge was how to make each of those devices accessible to any other. The approach used by the newly established Bell Telephone Company was actually borrowed from the solution devised by the telegraph companies. It was clearly impractical to run wires from every device to every other device. Instead these were brought to a central location,

Wires were everywhere, N.Y., 1909

which enabled the call to be switched to another similar device.

In the case of the telegraph, the usual method was to collect the incoming signal on a punched paper tape, which was then carried to the appropriate tape reader attached to the intended receiving device.

Lily Tomlin at work

As all those who have watched Lily Tomlin ("Are you the party to whom I am speaking?") will know, the primitive telephone switchboard operated in somewhat the same manner. A wire was brought from the device to a central switchboard. The connection was then made manually, by plugging that call into the line connected to the intended receiving device.

This of course did not solve all the problems. The photo on the next page shows the great telephone tower in

Stockholm in 1910, which was intended as a way of terminating the wires coming in from everywhere.

The picture below gives an idea of the overcrowding of the wires in a city even at the turn of the century.

The solution was to find a way to return to the preferred method of communicating, which is over-the-air without wires. That was not long in coming even though its ultimate rollout, cellular, would have to wait for almost 80 years.

The great Telephone Tower, Stockholm, 1910

City Wiring at the turn of the 19th Century

Cutting the Cord

In a fit of nationalistic fervour, I have subtitled this book *Canadians Cut The Cord.* That might be somewhat of an exaggeration, but Canada and Canadians did play a major role in the development of wireless communications and this continued right up until the development of cellular.

But before we get to the development of wireless devices, it is worth acknowledging the theoretical contribution of James Clerk Maxwell, who was definitely not a Canadian. He was a Scot, born in Edinburgh in 1831, and is sometimes ranked with Isaac Newton for his early fundamental contributions to science.

James Clerk Maxwell

For our purposes it is only worth noting that Maxwell's Wonderful Equations, of which there were four, laid the theoretical groundwork for over-the-air transmission. Heinrich Hertz, in 1887, discovered the electromagnetic radiation predicted by Maxwell in the latter's Treatise on Electricity and Magnetism of 1873. This electromagnetic radiation was what was needed to implement wireless communication.

$$\begin{aligned} \nabla \cdot \mathbf{E} &= 4\pi\rho \\ \nabla \times \mathbf{E} &= -\frac{1}{c}\frac{\partial \mathbf{B}}{\partial t} \\ \nabla \cdot \mathbf{B} &= 0 \\ \nabla \times \mathbf{B} &= \frac{4\pi}{c}\mathbf{J} + \frac{1}{c}\frac{\partial \mathbf{E}}{\partial t} \end{aligned}$$

Maxwell's Equations

Hertz was not interested, however, in the commercial exploitation of Maxwell's equations. He left this to Nikola Tesla, Guglielmo Marconi and Reginald Fessenden to demonstrate the practicality of this.

Tesla, who was born in Croatia in 1856, was the first out of the block. He demonstrated his wireless system in 1893 to the Franklin Institute in Philadelphia. It was not until 1895 that Marconi demonstrated his first system in Italy. There ensued over 30 years of legal battles between the two. It was not until 1943 that the U.S. courts struck down the Marconi patent in favour of Tesla's patent.

Nikola Tesla

Now to some of the Canadian connections. While Marconi was born in 1874 in Bologna and remained an Italian, dying in Rome in 1937, one of his most famous demonstrations of over-the-air transmission occurred on December 1901, when he received signals that had been transmitted across the Atlantic from Cornwall, England to what is now called Signal Rock at St. John's, Newfoundland.

Marconi was subsequently honoured at a dinner at the National Club on Bay Street in 1902, and the main floor reception area is known as the Marconi Lounge. (I was

President of the National Club in 1998–99.) Okay, this is stretching it but, at least, there is some Canadian connection.

However, Marconi had no interest in voice transmission. He felt that Morse Code was quite adequate for communication between ships and across oceans. Just as I and many others did not foresee the follow-on applications and uses of cellular, Marconi did not anticipate the development of radio and the broadcasting industry. For these reasons, he left the early experiments in wireless telephony to others.

Guglielmo Marconi

While there might have been some dispute about how 'Canadian' Alexander Graham Bell was, there is no dispute about the individual who really put wireless telephony on the map (or more correctly over-the-air). This was Reginald Aubrey Fessenden. Fessenden was born in 1866 in Milton, Quebec. He attended Trinity College School in Port Hope and then Bishop's College in Lennoxville, Quebec. His illustrious career started at the Thomas Edison Machine Works, where he became the Chief Chemist of the Edison Laboratory at Orange, New Jersey. In 1890 he became Chief Electrician at Westinghouse.

Prof. Reginald Aubrey Fessenden

Fessenden felt he could transmit and receive by a continuous wave method, rather than the spark-oriented apparatus that had been used prior to that time. The latter introduced substantial noise into the system and made voice communication over the air a very questionable venture.

In November 1906, Fessenden and his colleagues were perfecting their wireless technology by testing transmissions between Brant Rock and Plymouth in Maryland. To everyone's astonishment, his voice, telling the operator at Plymouth how to run the dynamo, was heard at Macrihanish, Scotland. It was with such clarity there was no doubt who the speaker was.

Fessenden's next triumph was on December 24, 1906 when he and his assistants presented the world's first radio broadcast. Fessenden actually played Handel's Largo on the violin. The broadcast was heard by ships at sea at various distances over the north Atlantic and as far south as the West Indies.

Wireless broadcasting had been born.

Fessenden's technique was Amplitude Modulation, best known in the form of AM radio.

To round out the Fessenden story, he has also been credited with inventing the radio compass, the sonic depth finder (SONAR), submarine signalling devices and even the turbo electric drive for battleships. He held some 230 patents and can clearly be called the father of AM Radio and the one who laid the groundwork for usable mobile telephony.

He died in 1932 in Hamilton, Bermuda, and despite his accomplishments, remains relatively out of the public eye compared to Bell, Marconi or even Tesla.

But all of this was still over the air from fixed locations. Mobile communication was still in the future.

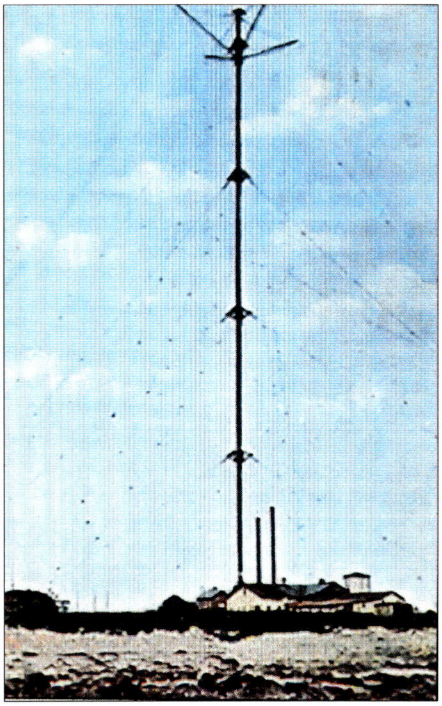

Fessenden's 420-foot transmitting tower at Brant Rock

An Early Wireless Phone Device

Annual Report
Smithsonian Institution
1908

WIRELESS TELEPHONY

by R. A. Fessenden

A Paper Presented at the
Twenty-fifth Annual Convention of the
American Institute of Electrical Engineers,
Atlantic City, N. J., June 29, 1908

reprinted by Lindsay Publications Inc

WIRELESS ON THE WAVES

Wires across the waves were obviously out of the question. One of the first serious needs for communicating between moving locations was for ships at sea. Given the limited power of early wireless telegraph, the antennae were initially somewhat cumbersome.

The photo of a battleship equipped with a wireless telegraph tower gives you some idea of what the early attempts looked like. This is from an interesting old book called *Wireless Telephone and Wireless Telegraph* by Charles Ashley and Charles Hayward published in 1912. However, the photo is clearly much earlier.

Battleships supplied with Wireless Telegraph equipment

It was not too long before it was determined that a wire strung between masts would do a credible job. It was this type of arrangement that was used for early marine telegraphy, the most dramatic example being the events surrounding the sinking of the *Titanic.*

While the mast shown on this early battleship looks ridiculous, we should bear in mind that these 'castles of steel' already had tall lookout towers right through the First World War. The reason was obvious. Without the benefit of spotter aircraft, GPS satellite systems or radar-directed guns, one needed to be as high as possible above the water to see where you were firing. The large guns on even relatively early warships could hurl a shell 20 miles, and this made spotting where the shell would land quite a challenge.

USS New York, May 1915

PHONY PHONES

It would be some time before a car phone would be a reality. However, it did not take people long to attempt to develop a car-based system of telephony. The photo shows an early model developed by the Marconi company in 1901 – it must have been murder taking it through the car wash!

The First Car Phone?

The search for a usable mobile phone that could carry voice signals to and from a vehicle attracted some very interesting early attempts. The financial potential attracted its own collection of fraudsters as well.

Anyone with the name of Nathan B. Stubblefield sounds suspicious. Nathan was born in Murray, Kentucky in 1860, the son of a prosperous lawyer. Without apparently much knowledge of earlier wireless experiments he developed an 'electromagnetic induction wireless telephone'. By 1902, he conducted a well-publicized public demonstration of the telephone and established The Wireless Telephone Company of America. It turned out that the demonstrations were fraudulent. His financial backers sued him, his children sold the family farm and his wife abandoned him. He became an eccentric hermit and died in 1928 of starvation.

Nathan B. Stubblefield

This did not stop Murray, Kentucky from claiming that it was the birthplace of radio in the United States. After World War II, they established their first local radio station with the call letters WNBS for Nathan B. Stubblefield.

More fun was to come.

A. Frederick Collins vacillated somewhere between being a technical genius and a fraud artist.

He allegedly developed a wireless voice telephone and formed a company called Collins Wireless Telephone Co. He became associated with some businessmen of questionable character, who proceeded to make wild claims about the technology, even predicting the downfall of telegraph stocks such as Marconi.

A. Frederick Collins – a Tragic Genius?

Their usual way of demonstrating the wireless system was between adjoining hotel rooms. What was not known to the prospects was that a wire carrying the signal actually ran between the rooms.

The Collins stock certificate is a work of art and like many stock ventures before and since, turned out to be completely worthless. In December 1911, the four officers of what was then known as Continental Wireless Telephone and Telegraph were indicted for fraud.

It is interesting to note the article in the Seattle *Sunday Times* of September 1909 showing what Collins claimed to be a “phone that will eliminate many of the troubles experienced while motoring at a distance from a garage.” As you will note, the gentleman with the phone is simply holding a standard land line phone. Collins was sentenced to three years in jail in Atlanta. When released, he went on to write a number of books including *The Radio Amateur's Handbook*, which is apparently still in print in its 82nd edition! He had a slightly better end to his life than Stubblefield.

The Collins Wireless Telephone Company Stock Certificate 16 April 1909

However, serious attempts at developing mobile voice communication were also underway. There was a real need for vehicular communication for emergency services, taxicabs and general business use. Before we look at some of these developments, it is worth spending a moment on the Electromagnetic Spectrum.

THE SEATTLE SUNDAY TIMES

SEPT. 5, 1909.

FOR AUTOMOBILES

The Collins Wireless 'Phone Will Eliminate Many of the Troubles Experienced While Motoring at a Distance from a Garage.

"Message from an automobile" – Seattle Sunday Times, September 1909

The Spectre of Spectrum

The electromagnetic waves that are used to communicate most information about the universe were described in James Clerk Maxwell's wonderful equations. Most of us have seen charts on the walls of our science classes showing the incredible range of such radiation. You may recall that the spectrum is normally ranked by frequency, e.g., how many peaks and valleys there are making up the wavelength in that particular form of radiation. This is important because the more dense the frequency, the more information the radiation can carry.

The spectrum ranges from very high frequency waves starting with gamma radiation through x-ray and ultimately into the ultraviolet area.

In the middle (more or less) is a very narrow band of visible light. This is all we humans can detect with our eyes.

Beyond that, with the wavelength getting progressively longer, are infrared waves, radar waves, microwaves and the ones that are used to carry television and then radio.

As an aside, evolution has ensured that our eyes can only detect the narrow range we call the visible spectrum. If we were able to see other parts of the spectrum, the visible world would be one indistinguishable blast of sensation. Remember Geordi with his visor in *Star Trek: The Next Generation.* He was able to visually detect radiation outside the normal visible light range.

The reason for mentioning this is that while there is a great deal of spectrum, only some of it is really useful for mobile or wireless communication. It is in relatively short supply and, as we shall see, very quickly became overcrowded.

As things in short supply have high value, the chase for spectrum started in the very early days of mobile communications and continues to this day.

So while there is a great deal of spectrum in total, the amount that can be conveniently used for wireless voice communication is limited. Large blocks were quickly reserved for specific uses, e.g., the aforementioned AM radio occupies the medium frequency area from 535 kHz to 1705 kHz (the unit of measurement called the Hertz is just a way of characterizing radio waves by their frequency. A kHz, or kilohertz, for example, represents

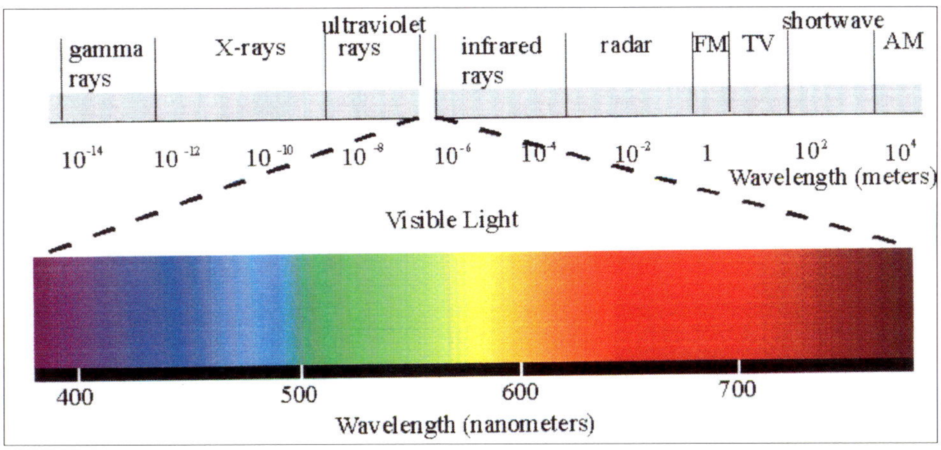

The Visible Spectrum – a small part of the Electromagnetic Spectrum

a thousand waves passing a point in one second).

Shortwave radio broadcasting is in the high frequency area and is measured in MHz (megahertz).

Many allocations were not made until later, e.g., over-the-air television channels that are in the VHF, or very high frequency (30 MHz) area, and a special allocation of channels that was hardly used for TV in the ultra high frequency, or UHF (300 MHz) range. I mention the latter because some of you who are older will remember the dial on the original TV sets that enabled you to access the over-the-air channels 14–69. It was this set of frequencies that was ultimately granted to cellular in its early days.

Many other frequencies were taken up for industrial, scientific and medical uses, or such needs as air traffic control, vessel traffic movement and the gigahertz range (1000 MHz) is where you find much of the satellite transmission. A small part of the spectrum in the UHF area (806–890 MHz) was allocated for trunked mobile radio services that we will describe in a moment and a bit in the same general area for paging. It was to make more efficient use of the spectrum that led to the birth of cellular.

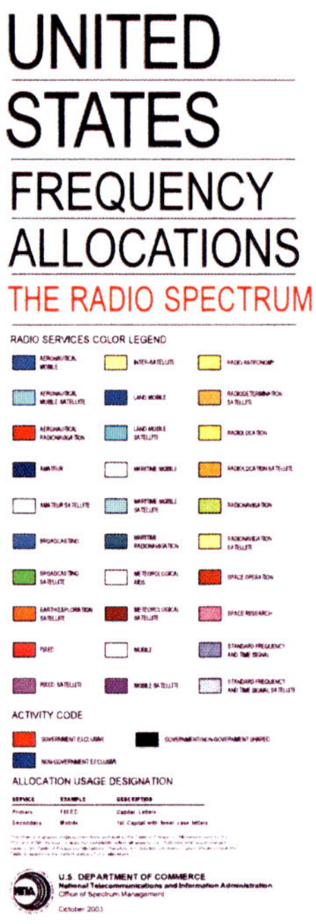

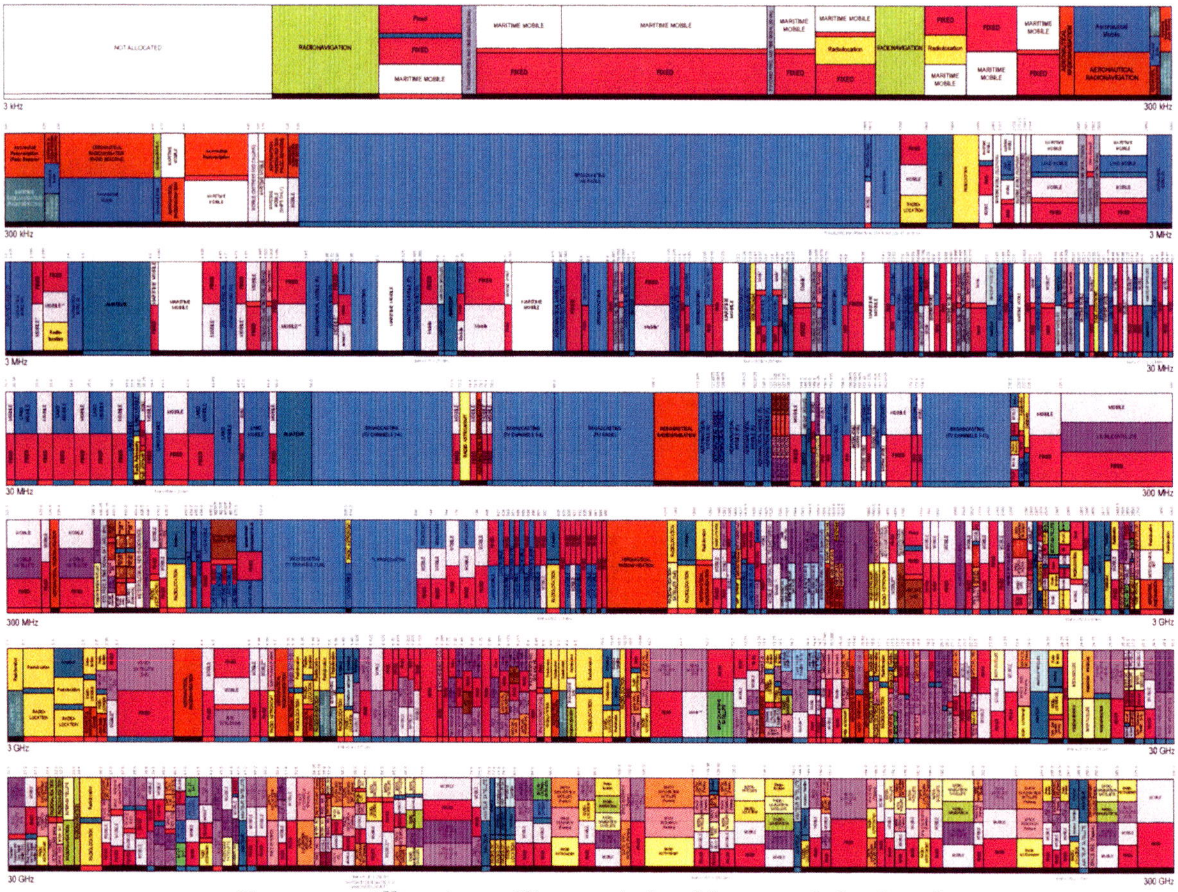

Frequency Allocation – You won't be able to read the details but will quickly see why Spectrum is so valuable

Mobile on the Move

As might be expected, the police were the first to try to exploit telephones in cars. It was the Detroit police department that first used mobile radios in their patrol cars in 1921. The initial transmission was one-way only.

Two-way communication appeared in 1933 when the New Jersey police department employed the first two-way, push-to-talk system.

The push-to-talk systems operated on a limited number of channels and these had to be shared. As the name implies, if you wanted to talk to someone on a channel, you had to push a button on the side of the unit in order to do so. There was no concept of a dial-up system in these early devices.

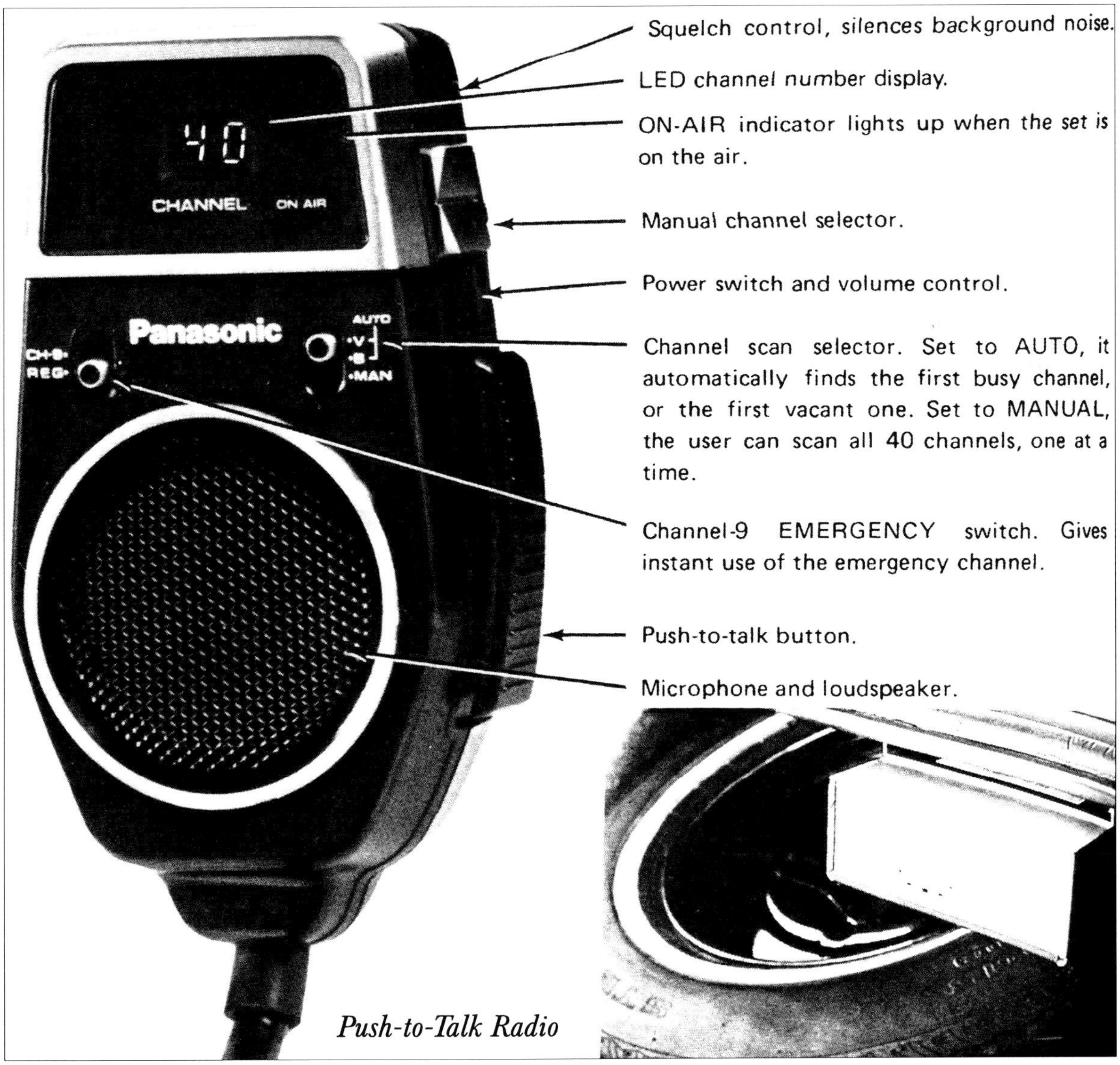

Push-to-Talk Radio

Caroline Van Hasselt, in chapter 12 of her aforementioned book, outlines some of the pioneering work done by Canadians in developing the first truly portable two-way communicators. Motorola, however, got most of the credit for coming out with its Handie-Talkie in 1940 for the U.S. Army Signal Corps. In 1943, Motorola developed a two-way FM portable radio called a Walkie-Talkie with a range of 20–35 kilometres. The Walkie-Talkie was a boon to the battlefield and formed the basis for the first hand-held cellular phone.

Dick Tracy Two-Way Wrist Radio

In the meantime the general public latched on to the idea of mobile voice communication with the help of Dick Tracy's 2-Way Wrist Radio. This did not become a reality for reasons we will get to later. Of course the technology at the time was not up to this.

AT&T's Bell system began offering mobile phone service for automobiles in 25 cities using a Motorola-built push-to-talk technology. However, this early radio-phone system used a single powerful antenna with only 25 channels. The service area was extremely limited and the number of users very restricted.

We talked about the efficient use of radio spectrum. In these systems each conversation tied up one whole channel on which one could transmit or receive. Caroline notes that in New York the system could handle just 12 simultaneous calls and the sound quality was horrible.

Bell can take credit for at least providing this type of early mobile phone system. However, it is understandable that with their experience on such a limited system they did not then really see the potential for a much more sophisticated system, with the ability to handle thousands of simultaneous calls.

It was only in 1970 that the Federal Communications Commission (FCC) agreed to allocate 75 MHz in the 800 MHz region for broader mobile use. There was little interest in making use of this.

Only in 1977 were two experimental cellular systems implemented, one in Chicago and one in the Baltimore/Washington DC area. By then countries in Europe and elsewhere got the lead in developing commercial cellular networks.

So how does cellular make more efficient use of the scarce radio frequency?

WHY CELLULAR SELLS

The concept is simple although, until the advent of sophisticated telephone switches, the technology would not have permitted widespread use.

Instead of having a single powerful antenna servicing a large physical area, that area is conceptually divided into a number of hexagonal zones or cells (see illustration). A low-power, low-height transmitter is set up more or less in the centre of each cell. The advantage is that the frequencies used in one cell can then be reused, not obviously in the adjacent cell or there would be interference, but in a cell farther away.

Each of the cell sites would then be linked to a Mobile Telephone Switching Office (MTSO), which would then redirect the call to a conventional telephone or to another mobile. There are a couple of additional layers, e.g., the MTSO would have to interface with a telephone toll office switch to connect to land line phones.

A further requirement was that as one drove through a cell and came to the edge, the call had to be handed off without interruption or interference to the next cell and so on as the driver continued to move from cell to cell.

In sparsely populated areas, cells might be eight miles across. In dense urban

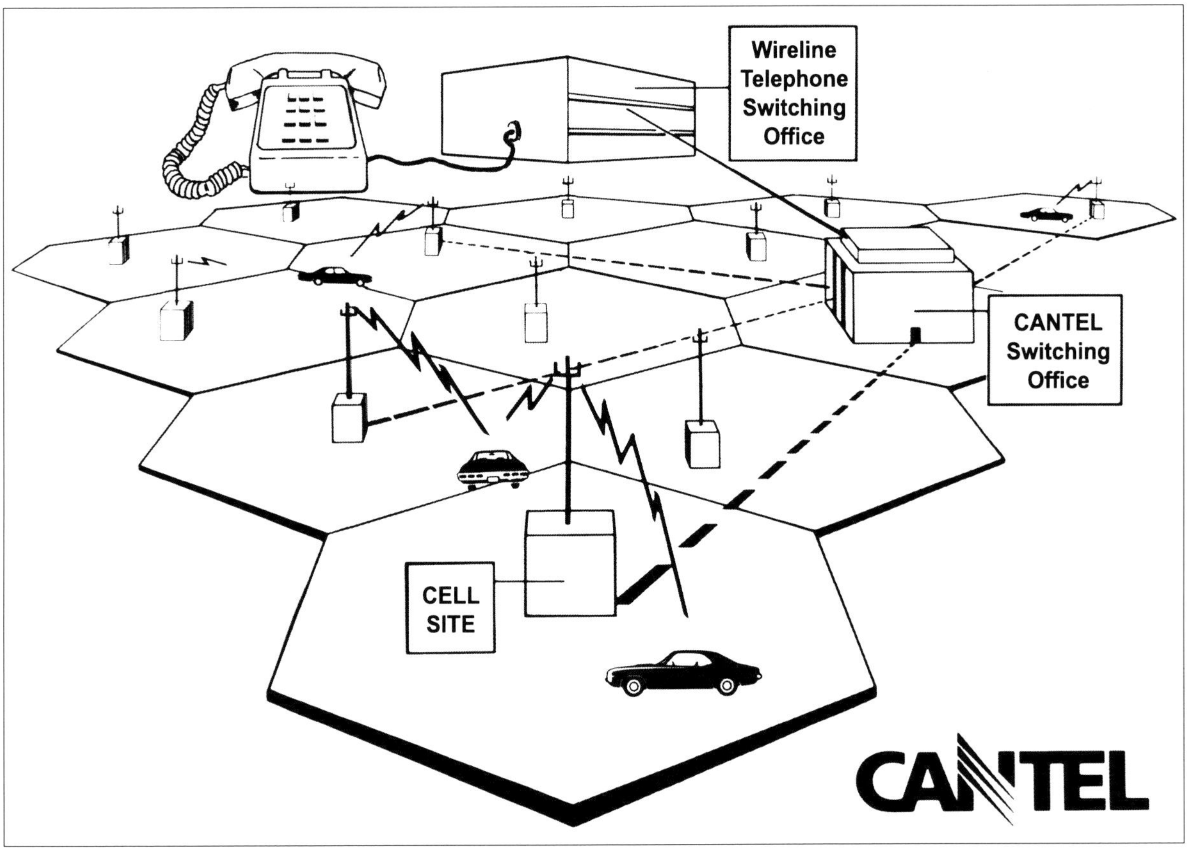

Diagram of a Cellular System

areas, however, they would be very close together and so handoff would occur many times.

The interconnection of the cell sites to the MTSO could occur over land lines or coaxial cable, such as is used for cable TV. A more common method, however, was to link these by microwave to the MTSO.

It is interesting that the idea for cellular likely originated at Bell Laboratories in 1947, where a D.H. Ring put forward a proposal that described how the system could work. It took until 1975 before the FCC gave final approval for Bell to start a trial system and another two years followed before it was allowed to operate. The first recognized cellular telephone call was made by a Toronto-born inventor, Martin Cooper. He made the call on a hand-held portable (about the size of a Kleenex® box), and allegedly made it to his competitor Joel Engel, head of research at Bell Laboratories. Marty was to play a role in the CANTEL application some years later.

The first commercial mobile phone system based on cellular principles was actually launched in Scandinavia in 1979, using the Nordic Mobile Telephone (NMT) system. However, it was the Advanced Mobile Phone System (AMPS) launched in the United States that would be the leading North American system for many years.

Cellular sells because the new concept provides the capacity to make it a viable consumer service.

Marty Cooper – helped with the CANTEL Application

CELLULAR SIZE

Canada chose to wait until the United States chose a technology – the AMPS system. They had little choice, as it was clear that customers would want the ability to use their phones anywhere in North America.

AMPS was an analogue system, which was not an efficient use of the spectrum. However, in the early days when usage was low, it was adequate for the task. It was a circuited switched technology, which simply means a cellular channel was tied up for each user, regardless of how intensely the channel was used to carry a voice signal.

It was also presumed that the initial major market for cellular phones would be in cars. This was not so much a reflection of how cellular providers thought the phones would be best used, but rather a reflection of the technology. The early phones were clunky. The first truly mobile phone came from Motorola, it was modelled on the Walkie-Talkies of World War II and weighed about 28 ounces. At 13 inches long, it was huge by today's standards. It was referred to as the Brick.

The cost for this unit was initially nearly US$4,000 and it provided just one hour of talk time and eight hours of standby time. The official Motorola name for the Brick was the DynaTAC.

The DynaTAC – not exactly pocket size

As this was hardly user friendly, this explains the early emphasis on cellular phones was for in-car use. For the sake of completeness, I should note there was a third type of cellular phone – the Transportable. As you will see from the picture, this involved a large and heavy battery unit (about 11 pounds), which had to be carried separately from the handset. While cumbersome, it did provide a substantially longer talk time because of its larger battery but, more importantly, it was a full 3-watt power unit. A word about this side of the technology is in order.

Because a cellular phone receives calls from a cell site that is a powerful stationary transmit/receive unit, there is usually little problem with picking up a signal. However, transmitting the signal from a much smaller hand-held or in-the-car unit is trickier. The power in a unit that you hold to your ear must be kept in a safe range. For hand-helds, this is 0.6 of a watt. For an in-the-car phone, however, where the transmitting unit is not close to your ear, the power could be substantially greater, e.g., 3 watts. The Transportable provided 3-watt power and hence was useful in areas such as construction sites.

This was the state of the technology used in the start-up systems in the United States, and would be where we were at when Canada decided to ask for applications for cellular providers.

As it was the Department of Communications that controlled and allocated radio frequencies, it was DOC who put out the call for tenders.

The Transportable with its large battery pack

Background to the Bidding

In May, 1981, the FCC in the United States decided to split the block of spectrum designated for cellular into two equal tranches, one to go to the local telephone company (wireline) and the other to a non-telephone (non-wireline) company. Canada chose to follow the same route.

June 7, 1982 was the deadline for Round 1 applications in the United States and, as might be expected, the FCC staff was inundated with filings.

On October 15, 1982, the DOC in Canada followed suit and called for licence applications for 23 cities. The applicants would have only 4½ months to pull together their bidding consortia, as the deadline for submissions was February 28, 1983.

The local wireline telephone companies knew they would get half the frequencies, but the upstart non-wireline bidders were going to have to compete for the other block.

The competitors quickly lined up for the non-wireline frequencies. I cannot improve on Caroline Van Hasselt's compact summary of the players and therefore will quote directly from her book: "Global TV's then co-owner, Seymour Epstein, and Toronto cable pioneer Geoff Conway joined forces with Western Union Telegraph Co. of New York to form Honeycomb Telephone Corp. Bell Canada's long-time telegraph and long-distance foe CNCP Communications teamed with U.S. cellular phone maker Motorola Inc. Winnipeg-based InterCity Gas Corp. partnered with the Toronto Bronfman family's merchant bank,

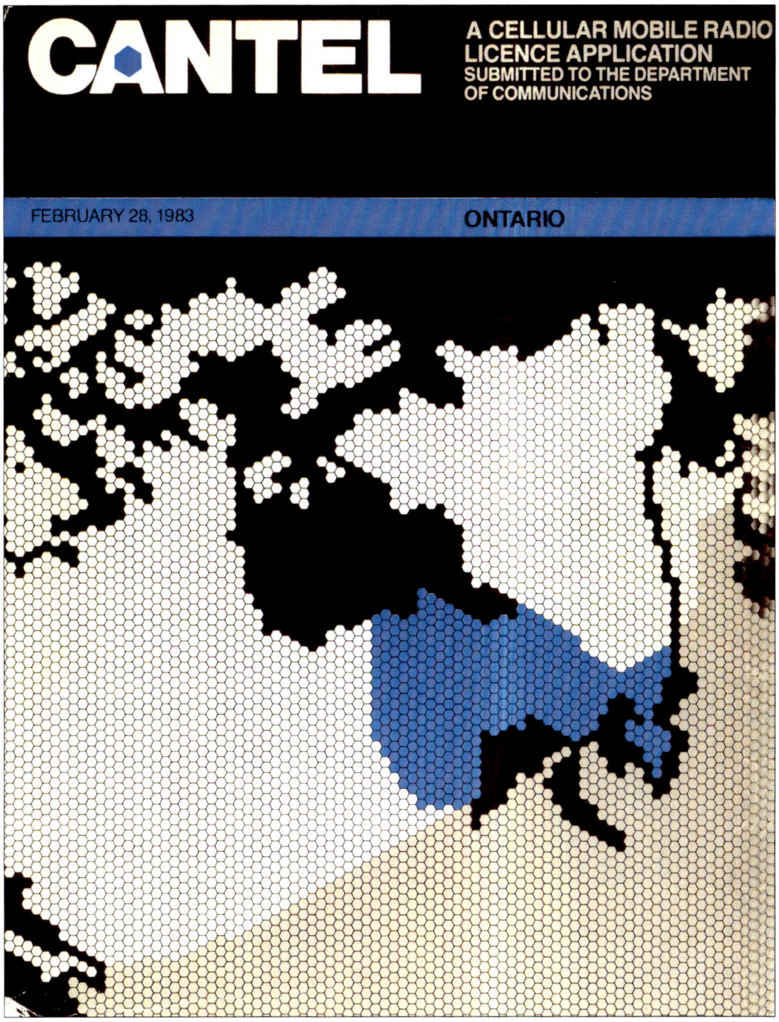

Cover of one volume of the 17 volume CANTEL submission, February 28, 1983

Hees International Corp. and the rebel Toronto securities firm, Gordon Securities Ltd. (the now defunct Gordon Capital Corp.). InterCity, which was aligned with New York-based Millecom Limited, called its consortium Roam Communications. Millecom had recently been awarded one of the two U.K. national licences and was vying for U.S. licences. Radio and TV network owner Selkirk Communications, and pulp and paper company Abitibi-Price teamed with Cellular Canada Communications Inc., a consortium of radio common carriers (RCCs)".

It is worth explaining that the RCCs were usually small companies that endeavoured to compete with the wireline companies in an early form of mobile voice communication. These companies provided a local push-to-talk mobile car phone service, and in some cases were also involved in the paging business.

The telephone companies were in this area as well, although other than Alberta Government Telephones (AGT) they did little to exploit the mobile phone market. AGT accounted for roughly half of Canada's then mobile phone market, principally because of the oil patch.

Shelley Kideckel was an entrepreneur in the Toronto area and won a court case against Bell to enable his company to connect his phones to the wireline network. It was an important decision, something similar to the CarterFone decision of June 1968 in the United States, which allowed non-wireline companies to connect their equipment to the monopoly telephone network.

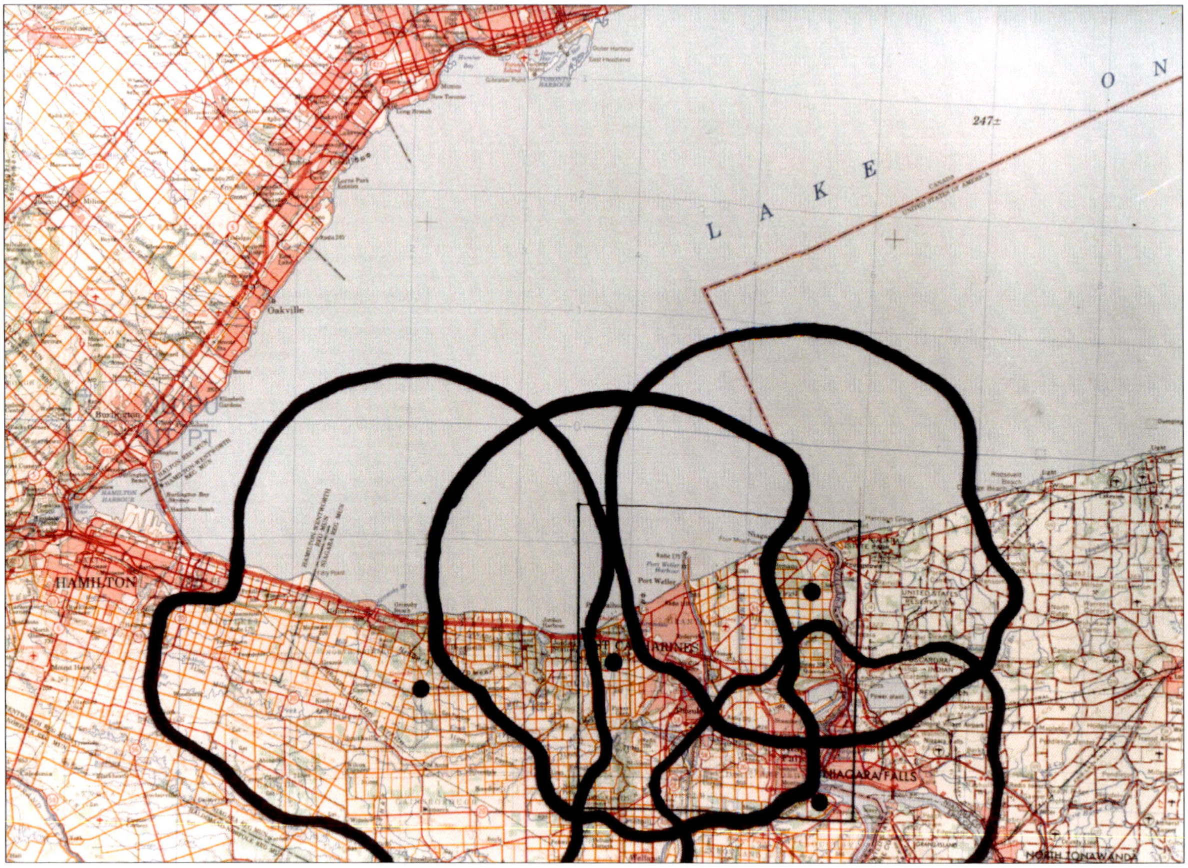

CANTEL's proposed coverage of the Niagara/St. Catharines area, February 28, 1983

Now, back to the Canadian bidding process.

There were two RCC bidders who, from the get-go, had announced they were only going to bid for frequencies in a local area in which they already had a presence. The first was CelTel in the Golden Horseshoe area, whose principals were Mike Kedar and Harley Murphy, and Time Cellular in the Ottawa area, whose principal owner was Botho von Hempeln.

These two RCC bidders clearly understood the local marketplace and could make strong regional bids. The DOC initially indicated that any company could bid for any or all of the licences.

The final entrant was by a company to be incorporated under the name CANTEL Mobile Radio Communications Inc. As Caroline points out, "It was backed by three of the unlikeliest partners, each having stumbled into this new uncharted world of cellular radio communications purely by happenstance."

She goes on to point out that this would become the most politically charged licencing process that Canada had ever seen.

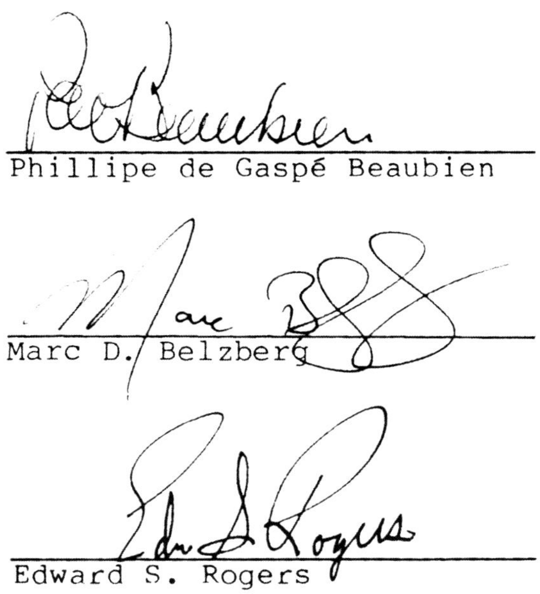

The Founders sign the original Memorandum of Understanding, February 25, 1983 – 3 days before filing!

BIZARRO

Ironically, the pedestrian he hit because he was on his cell phone was the guy he was on the phone with.
ANCE
BIZARROCOMIC.BLOGSPOT.COM
BIZARRO.COM
10-23-09

PART TWO

Let the Games Begin

THE COMING OF CANTEL

Overview

As promised in the Introduction, I am not going to duplicate the very good coverage given by Caroline Van Hasselt in *High Wire Act.* Section III of her book entitled Wireless gives many of the fascinating details about the financial gyrations and ownership issues that plagued CANTEL in its early days. As one of those interviewed by Caroline for the book, I can attest that it is surprisingly accurate and insightful for someone who was not actually involved.

My approach will be to include enough of the story that the reader of this book can follow the action but will concentrate on the interaction of the many people involved in a start-up that likely should never have started.

A relaxed photo of me in my then role as President and CEO of Premier Communications – flanked by Sandy Buschau, Director, Public Relations and Lal Fox, Senior Vice President

The Players

My involvement began at about the time of the filing of the initial replies to the DOC tender. The whole thing was typical Ted.

I was in Toronto for a meeting, although still living in Vancouver, and Ted asked me to come into his office. He inquired if I had any interest in assisting a group in obtaining the licences for cellular, given my prior enthusiasm for the project.

That interest arose from my being alerted to the forthcoming licencing process by King Margolese, a paging operator based in Vancouver. I had then discussed with Ted the possibility of Rogers bidding for the licences. King's proposal had been that his paging company would apply in Vancouver and he wanted to use the facilities of what was then Premier Cablevision to back-haul signals from the cell sites to the MTSO. I felt there was much greater potential and Rogers should at least consider this as a way of getting into the telephone business.

I had then sent Ted a section of a book by James Martin called *Future Developments in Telecommunications*, which outlined the technology and its potential. Ted had agreed as to the potential and had proceeded to look for partners.

Cover of James Martin's book

Having known Ted since the early 50s, I should have inquired more deeply as to what he meant by "providing some assistance." Ted advised me he had scheduled a conference call with his two other bid partners, Philippe de Gaspé Beaubien and Marc Belzberg. He asked me to stay in the room during the call but chose not to tell the others that I was there.

Ted opened by saying that while the applications looked good on paper, there was a huge amount of work to be done to convince DOC that we were not just a 'dark horse' but should be the front-runner. He then extolled my virtues as being someone who understood high tech (a bit of an overstatement, as I knew nothing more about telephones than how to push some of the buttons), was very familiar with the Ottawa regulatory scene (having lived there for 17 years), and was quite familiar with setting up new technology companies (having been the founding President of Systems Dimensions Limited in the computer services business).

What they really needed was a 'suit', and Ted said he would be willing to second me to the project to lead the team to get the licences. At that time I was Chairman of Canadian Cablesystems and what by that time was known as Rogers Cablesystems in Vancouver, Victoria, and with interests elsewhere in Canada and even in Ireland.

He proposed that I take over as President and CEO, and put together the necessary team to add credibility to the licence procedure.

This was starting to sound to me like more than just "providing some assistance"!

I had never met Marc, although I knew his father Sam and had met his mother Fran. Sam was born in Calgary, lived in Edmonton and had moved to Vancouver in 1968. He had formed a company called Western Realty, which was actually an amalgamation of 16 private companies.

Marc Belzberg

In 1970, Sam formed First City Financial Corp., which built up assets in excess of $5.4 billion, with operations throughout Canada and the United States.

I had known Sam through a variety of philanthropic projects in Vancouver, e.g., I had been Chair of the Finance Committee for the Board of Simon Fraser University and Sam became the Chairman of the major fundraising campaign about a year later.

Philippe de Gaspé Beaubien

I was also aware that Sam had something of the reputation of a 'greenmailer' but I found him a very delightful individual.

Philippe I knew extremely well. I am not sure that Ted realized how closely we had worked together. Philippe was the cousin of the late Ross LeMesurier of Wood Gundy, although neither could remember exactly what the relationship was. Ross had been instrumental in getting Wood Gundy to do an IPO for SDL in 1968/69. He recommended Philippe as a member of my initial SDL Board, along with Jim Tory (one of the founding partners of Tory, Tory, DesLauriers and Binnington), Walter McCarthy, Senior Vice President of Sun Life, and others.

Philippe and his delightful wife Nan-b became very close friends of mine. He was the one who recommended my entry into the Young Presidents Organization in Montreal, and was extremely helpful in the decade or so that SDL was a stand-alone company.

Philippe is the twelfth-generation scion of one of the oldest merchant families in Quebec. He had built a large communications empire called Télémédia and entered the broadcasting business in 1968. Télémédia owned a number of radio stations in Quebec and Ontario, and published magazines such as *TV Guide, Canadian Living, Harrowsmith, Country Life* and others.

He had been honourary chair of *PARTICIPACTION* and is perhaps best known as being the 'Mayor of Expo '67'.

Ted was a long-time friend and business colleague. We went through Trinity College at the University of Toronto together and were both Sigma Chi's.

The dynamics of the call were interesting. As might be expected, Philippe thought the idea of my taking over as President and CEO was an ideal choice. Marc was more hesitant. He acknowledged my credentials but I believe was harbouring a concern that I was 'Ted's guy'. However, there was no one else on the horizon and the partners agreed.

After Ted hung up, he turned to me and said, "That was easy."

It was about the last easy thing that would happen to me for the next decade!

The Game Begins

The DOC had received more than 50 applications, many from local paging companies.

Only two of the seven bidders, CNCP and Roam, submitted applications for all 23 cities. CANTEL had only submitted applications for 17.

Part of this was based on the initial ownership of the company. Télémédia would own 30% and would take the lead in setting up Quebec. First City Financial also owned 30% and would take the lead in western Canada. Rogers Telecommunications Limited, a late comer, owned 20% and would lead the charge in Ontario. The final 20% was set aside for the radio common carriers who, we all felt, would be an essential addition to the group, as they already occupied the territory with paging or primitive mobile telephone operations.

You will note that it was Rogers Telecommunications Limited (RTL) that owned the 20% and not Rogers Communications Inc., which was the public company. The story of Ted's overexposure to debt is well known. When the public company declined to participate in the bid, Ted took this on in the family company, although with the undertaking that if the public company wished to take it over at a later date, he would sell the ownership to RCI at cost.

This ownership arrangement meant the emphasis on applying for licences for CANTEL would understandably be in Quebec, Ontario, Alberta and British Columbia.

We now had another looming deadline and my first priority would be to put together a credible team to make a presentation of the proposal to the DOC on Tuesday, July 5, 1983. The presentations by the seven major bidders were to be spread over several days and we were given from 9.00 a.m. to noon. The locale was the Ottawa Congress Centre (the old train station).

Ted, George and Colin Watson, President, Rogers Cable after the successful sale of Premier to Rogers

The Team

Given the short lead time before the initial submissions on February 28, 1983, the 17 bilingual applications submitted by 'a company to be incorporated' had been put together largely by the Canada Consulting Group. The President was Jerome Redican who, together with his partner Sheelagh Whittaker, had done an unbelievable job of pulling together the material. They in turn used an American consultant, Kas Kalba, to provide the financial models, while Marty Goldfarb provided the consumer demand data.

The Canada Consulting Group offices, over the liquor store on Front Street where we developed the presentations

The applications included detailed proposed cell layouts and looked extremely professional – an amazing feat considering the time available to produce the material and get it translated. We were the only applicant to submit everything in both official languages.

All those involved in the initial application realized that Canadian content would be key, and our application contained the commitment to Westech Systems Limited to provide the cellular switching equipment. Westech, which was based in Calgary, was developing a fully distributed switching system architecture known as AURORA 800. It was, however, not much more than a paper system at the time we proposed to use it.

In fact, the switches from this company, which was later known as NovAtel, were never used by CANTEL.

We also proposed the use of General Electric base stations, as GE could supply them from Canada. This did not happen.

We undertook the development of the use of coaxial cable systems for inter-cellular trunking. This didn't happen either.

It was clear, as things rolled out, that we were going to have to maintain a very close relationship with the DOC to explain what was evolving, and it became obvious that I was going to be spending a lot of my time in Ottawa.

However, to get ready for July 5, I needed to pull together a team from whatever resources were available.

CANTEL

CANTEL CELLULAR RADIO GROUP INC.
LE GROUPE CANTEL DE RADIO CELLULAIRE INC.

EXECUTIVE COMMITTEE/COMITE EXECUTIF
PHILIPPE DE GASPE BEAUBIEN
MARC D. BELZBERG
EDWARD S. ROGERS

OFFICERS/DIRIGEANTS
PRESIDENT
GEORGE A. FIERHELLER
EXECUTIVE V.P. EXECUTIF
PIERRE L. MORRISSETTE
V.P. ENGINEERING/INGENIERIE
NICK HAMILTON-PIERCY
V.P. DEVELOPMENT/DEVELOPPEMENT
DAVID LINT
V.P. RCC OPS/RADIOCOMMUNICATIONS
KING MARGOLESE

SECRETARY/SECRETAIRE
CHARLES M. DALFEN

1818 CORNWALL AVENUE
VANCOUVER, B.C.
V6J 1C7
604-734-8411

P.O. BOX 2606
TORONTO-DOMINION CENTRE
TORONTO, ONTARIO
M5K 1J5
416-864-2373

1010 RUE SHERBROOKE OUEST
SUITE 1610
MONTREAL, QUEBEC
H3A 2R7
514-845-6291

A sample of our original letterhead

We needed names on the letterhead quickly.

Philippe agreed to lend me two of his senior staff.

Pierre L. Morrissette was Télémédia's CFO and he agreed to serve as the Executive Vice President and Chief Financial Officer of the new organization. Pierre went on to a stellar career in broadcasting after eventually leaving Télémédia. Amongst other things, the company he launched started the hugely successful Weather Network in Canada. Pierre continued to serve on the Board of CANTEL and its successor companies for many years and remains a close friend.

Philippe also agreed to provide the services of David Lint. David had come from McKinsey & Co. to join Télémédia as its Corporate Development Officer. It was David who got Philippe interested in cellular. He now joined the new team as Vice President, Development.

We needed an initial senior engineering presence. This was provided by Rogers, who seconded Nick Hamilton-Piercy, their Senior Vice President, Engineering. Nick was a great help as none of the rest of us had any telephone system expertise.

The aforementioned King Margolese became our Vice President, RCC Operations and gave us the link to the group that, in turn, had a strong link to DOC.

The late Chuck Dalfen – a close business associate and friend

The final name on the letterhead was the late Charles (Chuck) M. Dalfen. I had known Chuck from my cable days when he was Vice Chair of the CRTC. He had been retained by Philippe to assist in the cellular application. There could hardly have been a better choice. At the time Chuck was an independent lawyer and consultant based in Hull, Quebec. He had formed an association with Johnston & Buchan, a well respected legal firm in Ottawa. Bob Buchan was another Trinity type and we maintained an association for many years thereafter. Chuck became the Secretary of the company.

We had an Executive Committee formed of Philippe, Marc and Ted, and we showed addresses in Vancouver (mine at Rogers Cable), the Commercial Union Tower in the

Toronto Dominion Centre (Rogers' premises), and 1010 Rue Sherbrooke Ouest, which were the offices, at that time, for Télémédia.

All this looked quite national and surprisingly credible.

The Referee

The DOC team that would evaluate the proposals was headed by Vince Hill, Director General, National Telecommunications Branch. I would get to know and respect Vince who, on his retirement, I congratulated on being the true 'father' of cellular in Canada. He was the perfect example of the many wonderful civil servants we tend to under-appreciate in this country.

The Minister of Communications was the Honourable Francis Fox, who I did not know but came to know very well in the ensuing years. The Deputy Minister was Bob Rabinovitch, who I had known from my cable days.

I was also fortunate to know many of the Regional Directors. The DOC at that time was quite de-centralized. These Regional Directors would play a major role in the decision making.

The Honourable Francis Fox

So, on to the playing field.

The Pre-game Warm-up

My first face-to-face meeting with Marc took place at the Four Seasons Hotel in Toronto. A group of us sat around the pool area, where we felt we could have a private conversation. Chuck Dalfen was there and had changed into casual clothes. He had left his suit jacket in the locker room and I recall on his returning to get it, discovered his wallet had been stolen. It was not an auspicious start.

The purpose of the meeting was to plan the next step, with the emphasis on the all-important presentation on July 5. Chuck had spoken to Bob Rabinovitch, indicating that the July 5 date was not the best as it conflicted with the Telesat launchings in which Chuck and some others would normally have been involved. When I subsequently met with Rabinovitch, it was obvious he would be upset if he had to reassemble his panel members for some later date. I made the quick decision that we would go with the original date and so advised Vince Hill. Both men were very pleased that we were accommodating them.

In a memo from me to the Executive Committee, of June 23, 1983, I outlined the general plan for the presentation. I felt, as we were the least known of the bidders, it would be very important to have a large contingent at the presentation confirming the backing we had and our expertise in being able to meet the undertakings we had given.

Jim Green, President NovAtel Communications Ltd.

In particular, because of the emphasis on Canadian content, I wanted to ensure that Jim Green, the new President of NovAtel, would be there. I was quite pleased at the composition of the DOC panel. It was to be chaired by Vince Hill, and other members would include Richard Stursberg, Director General, Strategic Planning, who I knew quite well. As I had expected, the Regional Directors were all going to be there, and fortunately I knew Dave Lyon of Ontario Region, and John Quigley of Pacific Region very well from my cable days.

David Lint and I worked very closely with Jerome Redican and his associates from the Canada Consulting Group to structure the presentation. We virtually lived at their offices over an LCBO store next to the St. Lawrence Market.

I would be our chair at the presentation.

The others we assembled for the day included the three principals and, of course, all the officers as mentioned previously.

Jim Green was there and proved to be very helpful.

An RCC called Checkpoint Communications had agreed to join in our bid and their principals Bob Hill and Don Stewart were in attendance.

To bolster our technical expertise, Don Taylor, Vice President, Technical Operations of Rogers was there. As well, the consulting engineer who had done our cell layouts initially, Gordon Edwards, attended.

Bob Francis, Senior Vice President, Finance of Rogers lent his weight to the financing discussion.

Bob Buchan of Johnston & Buchan was there as legal counsel. We had Alan Flaherty of Kalba Bowen, and Peter Doering of Goldfarb and Associates, as well as Don Pagnutti, Vice President, Télémédia. Finally, as would be expected, Jerome Redican was there with one of his associates, Roger Martin. Roger would go on to even greater things as Dean of the Rotman School of Management at the University of Toronto.

Ted looked around the room and was delighted with the depth of the assembled group.

I led off by congratulating Vince on having selected a room with a hexagonal cell-shaped table, which apparently none of the other applicants had noticed. I assured Vince that not all the participants were going to speak, but I did have Nick and Don talking about the technology, David and King discussing our resources and our consumer service experience, Chuck to discuss the organization that we had proposed (quite decentralized but under the umbrella of the CANTEL Cellular Radio Group), and then Jim Green spoke about the benefits to Canada. Pierre Morrissette outlined the rather complex ownership structure.

I then took a calculated risk and put up a slide showing our comparative analysis of the seven principal applications. It obviously showed us leading in all areas. I saw Vince frowning a bit as he may have felt it was out of place for us to comment on others' applications. However, the information was all public and I later learned that DOC had come up with almost exactly the same chart and ranking.

The meeting concluded with brief presentations by each of the three principals, who spoke about their plans for their particular areas.

I had the feeling that we had done particularly well. In a memo to the Management Group and Executive Committee dated July 15, 1983, I outlined the next step that I recommended, as well as providing such intelligence as I was able to pick up. In that memo I noted that DOC had likely started into the process expecting that CNCP

COMPARATIVE ANALYSIS OF APPLICATIONS

	ORGANIZATIONAL					TECHNICAL		
	CANADIAN OWNERSHIP	FINANCIAL STRENGTH	CONSUMER SERVICE EXPERIENCE	TECHNOLOGY MANAGEMENT CAPABILITY	STRUCTURE	CANADIAN BENEFIT	PERFORMANCE/ COST EFFECTIVENESS	TIMELY START-UP
CANTEL	100%	Very Strong	Very Strong	Very Strong	Regional w/National Coordination	Maximum	Strong	12-18 months/ simultaneous
CELTEL	100%	?	Strong	Strong	Local	No Commitment	Weak	By 1985
TIME	100%	Strong	Strong	Strong	Local	Good	Moderate	In 1984
HONEYCOMB	80%	Weak	Strong	Strong	Regional	No Commitment	Moderate	15 months/ phased
CELLULAR CANADA	65-80%	Very Strong	Strong	Strong	Local with National Control	Very Good	Strong	?
R.O.A.M.	- about 60% of profits - 80.1% equity	?	Questionable	Not yet Demonstrated	National	Minimal	Not yet Demonstrated	?
CNCP/ MOTOROLA	51%	Very Strong	Very Strong	Very Strong	National w/Manufacturer Conflict of Interest	Conditional	Moderate	12-18 months/ Simultaneous

The Comparative Analysis Slide that nearly got us in trouble

would win. However, CNCP's involvement with Motorola, a U.S.-based firm, did not help and I learned their presentation had been embarrassingly poor.

The memo went on to outline my plans for setting up six teams to handle such things as dealing with Telcos (under Chuck Dalfen), procuring the right technology (under Don Taylor), planning for the organization and leadership (which I would chair), better understanding cellular economics (which Pierre took on), the pricing and service strategy (which David Lint would chair) and finally, continuing our constant interface with the government.

Vince subsequently asked that the written material provided at the hearings be put on the public record. We certainly did not object to this, although Francis Fox had evidently expressed some concern about our table comparing the various applicants. I hand delivered the letter on August 4, providing permission to Vince to make the material public.

On August 5, 1983, I wrote a memo to the Management Group and the Executive Committee that said: "I believe the decision has already been made. At this point, I believe it is in favour of CANTEL."

Vince and his staff were the perfect public servants and had given me no indication that we had moved from the back of the pack to being the front-runner. However, having been in sales since 1955, I had developed a sixth sense about finding out where we stood with a prospect. It came from interpreting the types of questions asked and

CANTEL

MEMORANDUM

CONFIDENTIAL

To: Philippe de Gaspe Beaubien
Marc D. Belzberg
Edward S. Rogers
P.L. Morrissette
N. Hamilton-Piercy
D. Lint
K. Margolese
C. Dalfen

From: G.A. Fierheller

Date: August 5, 1983

Subject: CANTEL PROGRESS REPORT

Front page of my memo of August 5, 1983 expressing the view that CANTEL was the front-runner

---

STATUS

As you know, DOC had requested that written material provided at the Hearings be put on the public record. Consistent with our previous position we have not objected to this, although we did point out the anomolies of the process as noted in the attached letter.

This letter was hand delivered to Vince Hill on August 4th. He clearly appreciated the flexibility we were providing to him as he also recognised and acknowledged the inequity of the position where we had put in extensive written material and others had put in very little.

However, it appears that the pressure for this move is coming from not only his legal department but also from Francis Fox. Evidently Fox expressed the concern, "If we award the licenses to CANTEL it might be construed that the competitive tables influenced DOC and others had not been allowed a chance to respond."

I believe the decision has already been made.

Although I caution you that this is only a gut feeling at this point, I believe it is in favour of CANTEL, although I have no way of knowing how many regions would be involved. I have some corroborating opinions from others close to DOC but nothing definitive.

Evidently, Francis Fox has been advised of the recommendation and it was this that led to the request the intensity of the discussion. Vince and his associates were not asking polite 'go away' questions, but were showing a level of interest that would only be taken if they felt this was the group with whom they would likely have to work.

But as it turned out, although my gut feel did prove to be correct, the bidding process was about to enter a second round.

The Rules of the Game Change

My optimism that CANTEL was on the verge of being the chosen supplier of non-wireline services in many of the markets was put on hold. On August 11, 1983, Francis Fox advised all bidders that DOC had decided it would be in the interest of Canada to choose a single supplier for all 23 metropolitan areas. This would ensure service to the smaller centres and avoid the likelihood of cherry picking just the larger markets.

Fortunately, CANTEL had advised Vince on March 1, 1983 that the CANTEL group would like to "build a national cellular radio network." We went on to point out in the letter that if the Department received either inadequate responses or no response at all to its invitation for licence applications to serve some of the smaller centres, CANTEL would be pleased to make further applications for these centres.

There was, however, an embarrassing problem now facing the DOC. Two of the seven finalists, Time Cellular of Ottawa and CelTel of the Golden Horseshoe, were clearly not in the position to bid beyond their local areas. Having expended a considerable amount of money on their applications, the new rules now eliminated them.

Realizing that this represented a major opportunity for CANTEL, I proposed to the Board that we immediately make an offer to take over these two applications in return for an equity position in CANTEL. The Board approved this approach and I was authorized to bring the two organizations into our application. We had set aside 20% of the ownership for radio common carriers and of this now offered CelTel 4% and Time Cellular 2.5%. They accepted.

Checkpoint Communications, based in Toronto, was an RCC already supporting our application as was Canadian Telecom Inc. (the Margolese family connection), based in Vancouver. We had now added powerful RCCs in each of Ottawa and Hamilton.

Our strong application was now significantly stronger.

The Canada Consulting Group did their usual fine job of pulling together the new applications and the submissions, once again fully bilingual, which were delivered to the DOC and the regional offices on October 14, 1983. Brent Belzberg and I delivered the package in Vancouver, in person.

There was little doubt at this time what our pitch was. While Time and CelTel added to our all-Canadian application, the other applicants had a significant foreign component. We pushed "Canadian ownership, a Canadian Board, Canadian engineering and Canadian technology, plus a Canadian sales and service operation."

Francis Fox felt it wise to have another oral presentation for each of the remaining five bidders. CANTEL was to be 'on deck' on Wednesday, November 16 at 9.30 a.m. in the DOC offices in Ottawa.

I wrote another memo to the Board dated November 9, 1983 called 'DOC – The Next Round'. To quote from that I said: "There is every indication that we are a front-runner and will use the occasion of this presentation to push the advantages of our application. It is not at the moment planned that we will spend a lot of time reviewing material on which we feel we are already strong. We have been asked to keep the group relatively small this time because of space limitations at DOC. I propose that we essentially limit this to Ted, Philippe, Marc, Chuck and me, and that the presentations be highly personal, with each of the original founders emphasizing their longtime commitment to Canada."

The cover of the second Submission, October 14, 1983

The story of that meeting has been well told elsewhere. The three founders did an outstanding job

• Marc spoke of how his grandfather had immigrated to Calgary in 1919 and how the family had started out in the second-hand furniture business

• Ted described his family background going back to the 19th century and the pioneering work of his father in battery-less radio

• Philippe noted that his forebears had travelled to French Canada with Radisson and Grosselliers in the 17th century.

Richard Stursberg quipped "that he was, regretfully, only a second-generation Canadian." Vince Hill added that it was "a slick presentation" although he later said it was not the personal appearances that confirmed the decision but rather the Application itself.

Richard Stursberg, then Assistant Deputy Minister Telecommunications, Research & Technology

Touchdown

The rest of the game is history. On December 14, 1983, Francis Fox called me to advise that CANTEL had won. This was followed by a Letter of Confirmation on January 3, 1984 to which I replied on January 9.

Peter Newman wrote in Maclean's (January 2, 1984) that this was "1983's most impressive business coup." It established Canada's first national telephone company.

Philippe phoned me the night of December 14 and said, "George, what do we do now? We are committed to building a national telephone system."

I jokingly said, "Not my problem, you just asked me to get the licences."

It did not work out quite that way.

TD

THE TORONTO-DOMINION BANK

Corporate Banking Division
Head Office
P.O. Box 1, Toronto-Dominion Centre
Toronto, Ontario M5K 1A2

Telephone No. (416) 866-5642

October 12, 1983.

Department of Communications
Government of Canada
Ottawa, Ontario.

Dear Sir:

RE: CANTEL GROUP

Cantel Group, which is comprised of First City Financial Corporation Limited, Rogers Telecommunications Limited and Telemedia Inc., are applying to you for licences for operation of cellular radio systems in various centres in Canada. We are advised that the partnership arrangements and participants differ in the various licence applications.

This letter will confirm that we are prepared to give favourable consideration to provide financing up to a total of $85,000,000. for these systems. This will be subject to negotiating satisfactory terms and conditions which could be influenced by the terms of licence.

The participants are all valued clients of our bank and we believe they have the resources to complete this project.

Yours truly,

Wendy A. Leaney
Assistant General Manager

Help from the TD Bank

wal/sf

Celebrating The Game Win

There is a strange sense of letdown mixed with the euphoria after any big win. The next day Roger Keay, who had joined the team from Rogers, and I went for lunch at a small restaurant in the basement of Commercial Union Tower. It was hardly a champagne & caviar event. We were both somewhat worn out from months of effort and could only imagine that the many others who had worked so hard to get us to this point were in the same state.

There was no big celebration.

The partners asked if I would stay on as President of the corporation, with the two objectives of pulling together more detailed plans for the launch and, most importantly from my standpoint, choosing a new President. I was still a Vancouver resident and was looking forward to my retirement in that area.

As I was also acting as Chairman, I arranged for an immediate meeting of the Board of Directors (the three founding partners) to be held at The Toronto Club on January 3 and 4.

At that meeting I noted the task forces had been somewhat rearranged using the seconded help from the various partners. Oliver Bush had now joined the group from First City and he chaired the financial task force. David Lint undertook Marketing; Nick Hamilton-Piercy, Technology and Engineering; Chuck Dalfen, Legal and Regulatory; and I undertook the Management and Corporate Communications role. The Board acknowledged that these task forces would be 'under the general direction of the President'.

Fox announces award of national cellular radio licence to CANTEL:
Decision marks start of new $800 million communications industry.

MONTREAL, December 14, 1983 – Communications Minister Francis Fox announced today that he will issue radio licences to CANTEL of Montreal to develop the world's first national cellular mobile radio telephone service in 23 locations across Canada.

Mr. Fox told reporters at a news conference in Montreal that: "This decision marks the start of a new communications industry for Canada. In the next three to four years, CANTEL and Canadian telephone companies will invest up to $500 million to establish national cellular radio services. An additional $300 million will be spent by users of the system for equipment. Canadian equipment and service suppliers in all regions will benefit from this investment."

"The potential for long-term return on investment is also extremely high," Mr. Fox said. "It is estimated that the cellular radio industry will generate gross revenues of $180 million per year by 1990, and we foresee even greater market expansion beyond that date. The potential for export is also phenomenal, with a world market estimated to reach $10 billion annually by 1990."

First page of the press release from the Honourable Francis Fox announcing our win

The other quick decisions were to retain a search firm for the permanent President – Massey Charbonneau and later Caldwell; select a lead bank – the Toronto Dominion, with the Royal Bank as an additional banker; and to solicit proposals for an auditor, from which we selected Price Waterhouse. Tory, Tory, DesLauriers and Binnington would be the counsel for financing and tax matters, and the firm of Johnston and Buchan (with Charles Dalfen) would be the general corporate and regulatory counsel.

We were on our way.

A happy-looking George

Chuck Dalfen, also looking happy

THE CELLULAR SERIES

We had won a game that no one expected us to win. Now we had to prove we could win the Series.

The Character Of The Characters

I mentioned that the founding partners proved to be both a help and a hindrance. It is worth looking at the motivations of each.

The partners had held together reasonably well during the bid process. But until we had the licences, there was no value that would give rise to disputes. All there had been to that point were costs.

Chuck Dalfen reminded me that at one point during the bid process I had asked, "Are you sure you really want to win this?" Perhaps Chuck and I were more aware than the others that once we had won licences for a national telephone company, we were stuck with huge financial obligations, to say nothing of the risk to our reputations if we failed to deliver.

Let's look at the principals, starting with Marc Belzberg.

Marc Belzberg

Although Marc always denied that it was his intent, his family, who were putting up the money, had a reputation for being 'flippers'. They knew that you cannot traffic in licences, i.e., win them and immediately sell them for a profit. However, I believe that Marc's ultimate motivation was really to build up as much value as possible, as quickly as possible, and then find a buyer. Marc had no experience in running organizations and I doubt if that was really his ambition.

Philippe, on the other hand, took his obligations very seriously. However, as it would play out, Philippe had numerous other calls on his money and would prove to have some difficulty financing his portion over the long term.

Ted was clearly a builder. His modus operandi was to select a technology in which he firmly believed and then borrow whatever was required to make the system successful. However, Ted had another trait. He was so determined to make a success of whatever he undertook that he had become a control freak. Although he often started with partners, to my knowledge he never once stayed with a partner for any length of time. He always noted, correctly, that all his partners made lots of money from the partnerships, but ultimately Ted only felt comfortable when he had 100% ownership of whatever the venture was.

CANTEL would prove to be no different.

It was not that Ted did not try to get along with partners, but his ideas of how to run an organization were unbendable and, eventually, he simply wore his partners down until they agreed to sell.

Not only were the motivations and methods of the three partners very different, but their backgrounds and personalities could hardly have been further apart.

Marc was Sam's only son and was considerably younger than either Philippe or Ted. After a somewhat directionless youth as Marc himself described it, he found religion and became an ultra-Orthodox Jew. He studied Hebrew and observed the Jewish dietary laws to the extreme. For example, he would not eat off any plate or use any utensil that might at any time have come in contact with non-Kosher cooking. It was a first for The Toronto Club, I am sure, when he insisted not only that Kosher food be brought in from outside, but that it be presented on plastic plates with plastic knives and forks that had never been used.

His approach to religion and life, however, seemed extreme to several of us. Although I was not there when he allegedly made the remark to Philippe that "religion is religion, business is business," this would have shocked me (as an atheist) as much as it did Philippe, who I believe one could describe as a non-practising Catholic. Neither of us could understand how anyone could be as professedly religious as Marc and not have these values translate into everyday business dealings.

Marc gave the impression of being suspicious of the motives of everyone else. Perhaps that was just because he was young, and concerned that his partners would somehow outmaneuver him.

Philippe, with what was an almost aristocratic French-Canadian background, was a true gentleman. A handshake closed a deal, not a complex Shareholder's Agreement.

Ted was clearly the entrepreneur, driven by the desire to put back together the family fortunes that had been so tragically dissipated through the early death of his father. He was the most driven person I have ever met. His point of view was pure WASP, with a total commitment to the Puritan work ethic.

Philippe and Nan-b de Gaspé Beaubien

So there you have it. A young, brash, ultra-Orthodox Jew from western Canada, an aristocratic, French-Canadian gentleman from Quebec, and a Protestant-ethic driven, extreme entrepreneur from Ontario. During the licence presentations, we had made something of a joke of this and it had, perhaps, played a role in our getting the licences.

It was now going to make life very difficult for whoever was the President. As noted, this job fell to me, at least for the interim.

Boards And Budgets

CANTEL had no financing and at that point did not even own a box of paper clips. The only staff was seconded from the three partners and these were paid for by the companies from which they were loaned.

My first request from the Board was to table a preliminary budget. It was for $1.5 million.

This had nothing to do with the network build of course. It was just to finance the most basic start-up costs including:

- initial space
- start-up secretarial assistance
- retention of a firm to develop the image, logo and other necessities
- outside consulting from Canada Systems Group and others for further development of marketing and technology plans
- legal, audit and other expenses
- most importantly, the fees for a search firm to find a President.

The partners expressed some shock but agreed to put up the money proportionally. This was done partially by guaranteeing a $1 million Line of Credit with the TD Bank.

It is hard to believe how little there was to work with. There was no permanent office. I simply borrowed a chair and desk from whoever was not using it at the time in the Commercial Union Tower space of RCI.

The entire corporate files were in a single legal size case that I took home with me at night.

Attack Ads of the type that the Competition never used, thankfully

We needed an immediate sense of permanence, which meant retaining some space and, at least, getting a telephone number and mailing address. Barry Ross, a Vice President of Rogers, who assisted Ted in real estate transactions, now assisted me in locating our first physical location at 20 Queen Street West, in the Cadillac Fairview Tower. It was not much space but it was a relatively inexpensive sub-lease that at least got us started.

We contracted with Burson-Marsteller to develop a media relations program, which I also felt was vital at this point. They won the bid ($40,000) by developing a series of mock ads that they felt might be used by the Telcos to attack our credibility. Some samples are shown on the previous page.

We also contracted with Don Watt & Associates. I had used a firm called Brake, MacDonald, Paine and Watt for developing the logo and graphic presentation for SDL. We now asked Don to do the same for CANTEL.

We made an early switch from Massey Charbonneau to the Caldwell Partners for the presidential search, and I estimated that this search would cost close to $200,000.

Commercial Union Tower – now Canadian Pacific Tower

20 Queen Street West – our first real office

A Man of Steel

Before we get into a brief review of the many challenges facing CANTEL in the first quarter of 1984, we need to address my top priority – finding a replacement for myself as President.

With the help of the Caldwell Partners, we identified our first choice as being John McLennan. I had known John somewhat from my Ottawa days when he, Terry Matthews and Michael Cowpland were building Mitel Corp. John had actually left Mitel and was doing some part-time consulting based in Ottawa. While John was really intrigued with the future of the cellular business, he was reluctant to move into full-time management and equally reluctant to relocate to Toronto, which would likely have been a requirement. On April 3, 1984, I reported to the Board that John had declined our offer.

The next candidate was Walter F. Steel. I did not know Walter, but he came highly recommended as a solid manager who had spent a decade with Bell & Howell and then took over as President of a struggling company called AES Data Inc. After successfully turning this company around, Walter had retired at 47. Perhaps a flag might have gone up at this point, as one of his hobbies was big game hunting.

Walter Steel, President & CEO

Walter was interviewed by the three principals and by me. Walter later commented to me that from the start he had a feeling Marc did not like him. I felt he had the right stuff to manage the company through its very difficult set-up stage. On April 27, 1984, I recommended to the Board that Walter be hired as President and CEO, and he accepted.

As Caroline Van Hasselt noted very accurately, "He was intrigued enough by the challenge of building another bleeding-edge company. It would become the worst experience of his previously stellar career."

At the same April 27 meeting, I was appointed Chairman. It was with a sigh of relief that I passed the reigns over to Walter.

Building The Team

While the search for the President was going on, I had to launch a parallel search for a permanent team to replace the interim seconded group. It was not going to be an easy job, as CANTEL was a start-up company with little telephone experience and no solid financing.

The most important single position was Vice President, Engineering. The candidates who were presented to me were less than stellar. Then, fortuitously, a Télémédia director, Stephen Kauser, mentioned to Philippe that his younger brother, Nick, might be interested in the job. Nick was a McGill University graduate, although at the time he was working in Venezuela with their national telephone company. He subsequently started his own business installing and servicing telecom equipment and networks. However, he really wanted to raise his family in Canada and was ready for a move.

Ted Rogers, Nick Hamilton-Piercy and I met with Nick for dinner at the King Edward Hotel. I outlined the challenge of building a major cellular system in less than a year. Ted asked, "Do you think you can do it?" Nick replied, "Do I have any choice?"

This was exactly the right response and, on the spot, Ted said, "Welcome to CANTEL."

I could not have been more pleased as I had hit if off with Nick right from the start. He was not only absolutely key for the subsequent success of CANTEL but became a close personal friend. I recall the first time I met Judy, his wife, when the Kausers and I were staying at the King Edward Hotel and the fire alarm went off. I was introduced to Judy standing more or less in our night clothes on the street in front of the hotel. Over the ensuing years, we would take several trips together.

It did not take Nick long to make up his mind. He spent a few days with Roger Keay reviewing the situation and then agreed to come on board.

It was very important that CANTEL negotiate with Bell to establish the relationship between the two companies. One of the many people I interfaced with was Joe Church who, at that time, was Assistant Vice President, Policy Development, for Bell. We badly needed someone who understood the regulatory scene at Bell to supplement the very good advice we were getting from Chuck Dalfen, Bob Buchan and others. We hired Joe as Vice President, Corporate Development.

Clearly Sales and Marketing would be vital and our next hire was Paul Kavanagh, who had many years' experience with Philips Information Systems, Xerox and others.

It was amusing that Sheldon Kideckel had called me, about two days after we were awarded the licence, to indicate that as he was the leading RCC in the Toronto area he could solve all my sales problems if we gave him an exclusive for the province of Ontario. I thanked him for his offer but politely declined. Shelley did go on to become one of our CANTEL Service Centres, as we will discuss in a moment.

The next hire was logically enough a Vice President, Finance and Administration. For this we chose David Perks, who had previously been the Chief Financial Officer of Hudson's Bay Mining and Smelting and had spent 12 years with the Globe and Mail.

Nick Kauser, Vice President Engineering

Joe Church, Vice President Corporate Development

Paul Kavanagh, Vice President Sales & Marketing

David Perks, Vice President Finance & Administration

Marc Ferland, Vice President & General Manager Eastern Region

With the emphasis on Quebec, I felt it was very important to have a Vice President and General Manager, Eastern Region. For this position we chose Marc Ferland, who had held positions with companies such as Canadian General Electric, and was recently General Manager of Compugroup Business Systems Limited.

With this team in place and Walter coming on board, the task forces that I had set up were now phased out.

Many, such as Nick Hamilton-Piercy, just went back to their regular positions. Roger Keay, however, elected to join CANTEL, reporting to Nick Kauser, and he proved to be a very helpful addition to the team.

David Lint saw himself as a candidate for the Vice President, Marketing job, but this was one position where Ted expressed some reluctance. David and Ted were very different personalities and probably would not have worked well together.

On April 30, 1984, I tabled a lengthy memo to the three founding partners that resulted from a discussion I had with Walter regarding the relative roles of the Chair and the President. There was nothing particularly surprising, and I had assured Walter that I would stay out of his way as he moved the organization forward for the launch. I would concentrate on the normal responsibilities of a Board Chair, establish an Audit Committee and the like and would work with Walter on the development of a strategic plan, government relations, legal and regulatory and would assist in major sales contacts (we had already started those with companies such as Canadian Tire, The Bay, Budget Rent-A-Car and others) and the general area of public relations.

We now needed to set up a more permanent Board of Directors.

Building The Board

In anticipation of some form of public financing, it was necessary to expand the size of the Board.

The three founders would remain as the Executive Committee. Walter and I were added to the new Board and each of three founding partners named two additional Board Members. Marc chose Sam Belzberg and Morty Cohen. Philippe picked Stephen Kauser and my long-time friend from the SDL Board, Ross LeMesurier. Ted picked his stepfather John Graham and John McLennan. Time Cellular appointed Zeev Vered, an Ottawa-based consulting engineer, and CelTel chose John Sheppard.

The final member was Chuck Dalfen, who continued as Secretary.

It was actually a very good initial Board and proved to be some calming influence on the Executive Committee.

But before Walter even arrived on the scene, there were a number of pressing priorities that had to be addressed.

Nick Kauser and me in Paris

PRIORITIZING THE PRIORITIES

Head Start

As noted earlier, the Telcos knew in advance they were getting half the frequencies and therefore had proceeded, at least in Ontario and Quebec, to put switches in place and prepare for an early launch. Bell had already stated its intention to start offering service in September 1984. In our proposal, we had pointed out to the Minister that 18 months would be required for the non-wireline to reasonably be expected to be ready. This would put the start date in mid-1985.

An early start by the Telcos would have been ruinous for CANTEL. At the second Board meeting on January 19, 1984 at the York Club in Toronto this issue was already on the table. Sheelagh Whittaker of The Canada Consulting Group had put together a series of options, and it was suggested that if Bell could not be delayed in its launch, CANTEL could start marketing on the Bell system. We would still obtain our own block of phone numbers so that the customers could be transferred to our system whenever it was ready.

This was undoubtedly a last-resort solution, and Chuck Dalfen and I immediately set about lobbying Ottawa to delay the start-up by Bell.

An article had appeared in *Fortune* on February 6, 1984 under the title Cutthroat, which outlined the experience in the United States where a head-start situation was allowed. I used this and other arguments with Vince Hill, and proceeded to get the support of Jim Green of NovAtel, Ray Doucet of Douserv Group Inc. of Montreal and others to start a letter campaign.

I pointed out that the problem as I saw it was not just with Bell. As we rolled out across the country, we would face the same head-start situation with each of the Telcos.

I summarized these arguments in a Telex to Minister Fox and on March 14, 1984, he announced that not only would Bell not be allowed to start until July 1, 1985, but that in each of the other Telcos' jurisdictions they would not be able to start until six months after a suitable interconnect agreement had been signed with CANTEL.

The Globe reported that the announcement "startled Bell Canada." Francis said in the same article that, "I think there would not have been a CANTEL in 10 months" if I had allowed a head start.

We now had the schedule we had asked for in our original application, but it was still extremely tight. At this point we had no bank financing, no contract with an equipment supplier, virtually no staff and not even a head office.

By mid-March Bell had already installed its cellular switch and main antenna in Toronto. We did not at that time even have a permanent CEO as Walter did not arrive until April.

CUTTHROAT

Competition in mobile phones

■ Not long ago, starry-eyed entrepreneurs were gazing at cellular technology and sighting a $10-billion mobile-phone industry in 1990, with heavenly pretax margins of 35% to 40%. Now the technology—which involves creating many radio-transmitting "cells" within cities to clear the jammed airways for car telephoning—is coming to market. The Federal Communications Commission has awarded construction permits in 38 large metropolitan areas to the local phone company (and its partners, if any). The FCC is also awarding permits to a so-called nonwireline operator in each city but so far has picked only ten of the winners.

Being first is a big advantage because there's great pent-up demand. In Chicago, which last October became the first city with full-scale cellular service, the local phone company, an Ameritech subsidiary, has the market to itself and is signing up eager customers at the unexpectedly high rate of 100 a day. So far 5,000 subscribers have been willing to pay the $50 monthly access charge, the 40 cents per air-minute peak rate, and the $2.500 to $3.000 for in-car equipment (telephone and antenna). Subscribers (mostly self-employed businessmen) are getting average monthly bills of $200.

The big question is how far competition will drive down those prices, so all eyes are on Indianapolis, where two companies are gearing up for battle. The "wireline" company is a venture led by GTE and Ameritech called GTE Mobilnet of Indianapolis. The competitor is Indianapolis Telephone Co., led by Graphic Scanning, a New Jersey-based telecommunications company (fiscal 1983 sales: $114 million). Graphic Scanning says it is a partner in companies that have won the nonwireline permits in all ten cities, though its stake is clouded by an FCC investigation of charges that the company falsified information in its applications to the commission. Now Indianapolis Telephone has filed a petition to stall the GTE-Ameritech system in its tracks. Charging that GTE Mobilnet violated its Indianapolis permit by changing the site of a radio tower, the Graphic group is asking the FCC to deny GTE Mobilnet a final go-ahead. Since the first operator can grab the eager-beaver customers, who also will presumably be the heaviest users of the service, GTE Mobilnet has a lot to lose. Says **Henry Lucas**, 40, who just left as chief of GTE's cellular operations, "Everybody wants those first customers. Everyone is saying, 'Pull out all the stops. Find any way you can to halt the other guy.' "

Kevin Blair, 35, who is handling Lucas's old job, claims that Indianapolis Telephone has resorted to "regulatory means as a competitive tool." Indianapolis Telephone's boss, **William Kokorelis**, 39, retorts, "I don't think it was dirty pool," and says he will be selling service any day.

Indianapolis Telephone plans to charge an access fee of only $30, plus a peak rate of 35 cents per minute for calls. GTE Mobilnet says its rates will be higher: a $33 access charge and 39 cents per minute. (The in-car hardware is made by independent manufacturers.) Kokorelis believes that in ten years both the service and the hardware will cost no more than regular phone service. Most experts, however, think that it will take at least five years to cut the cost of the customer's equipment by half and that prices for the service will also drop slowly. But cutthroat competition is the thing that could dim those astronomical profit projections, so Indianapolis will remain the city to watch.

STAFF *Eleanor Johnson Tracy, Steven Flax, Richard I. Kirkland Jr., Ann Dransfield, Tom Post, Monci Jo Williams*

Headstart article, Fortune, February 6, 1984

T
O
MARCH 7/84
TO EDWARD S ROGERS

THE FOLLOWING TELEX WAS SENT TO FRANCIS FOX XB TODAY

GEORGE FIERHELLER

MARCH 7, 1984 10:30A

TO: THE HONOURABLE FRANCIS FOX
MINISTER OF COMMUNICATIONS

WE ARE VERY CONCERNED TO SEE ONCE AGAIN IN THE GLOBE AND MAIL THIS MORNING A LARGE AD BY BELL CANADA INFORMING THE PUBLIC OF THEIR INTENTION TO START PROVIDING CELLULAR PHONE SERVICE IN MONTREAL AND TORONTO IN SEPTEMBER 1984.

BELL HAS NOT TO OUR KNOWLEDGE BEEN LICENSED BY DOC AND YET STATE THAT ''INSTALLATION IS WELL UNDERWAY IN TORONTO AND MONTREAL''.

WE UNDERSTAND THAT IT IS THE RECOMMENDATION OF DOC STAFF THAT THE PROCEDURE OUTLINED IN MY LETTER TO YOU DATED FEB 20, 1984 BE FOLLOWED.

I BELIEVE IT IS IMPERATIVE THAT A PUBLIC ANNOUNCEMENT BE MADE BY YOUR OFFICE AS SOON AS POSSIBLE IF YOU ACCEPT THE RECOMMENDATIONS IN XMM THAT LETTER.

WE WOULD HAVE TO SERIOUSLY RE-CONSIDER THE FEASIBILITY OF OUR STARTING AT ALL IF OUR COMPETITOR IS ALLOWED A HEADSTART IN ADVANCE OF OUR REALISTIC START DATE AS ACCEPTED BY DOC.

YOUR IMMEDIATE ATTENTION WOULD BE OF GREAT ASSISTANCE.

G.A. FIERHELLER,
CHAIRMAN,
CANTEL CELLULAR RADIO GROUP INC.

CC: PHILIPPE DE GASPE BEAUBIEN
MARC D. BELZBERG
EDWARD S. ROGERS

ROGERSCBLI TOR
065-23996+

My Telex to the Honourable Francis Fox March 7, 1984

Interconnection

This is a complex issue but the details are less important than the principle. As noted earlier, until very recently it was not even possible for a company to own its own phone equipment and interconnect it to the Telco network. The Kideckel decision changed this, but Bell was still proposing that CANTEL be treated as a customer rather than as a full-fledged telephone company. Needless to say we objected strenuously to this arguing that we had been licenced as a telephone company under the Railway Act (believe it or not this was still the Act that defined such things). CANTEL's position was that we should be allowed to have our switches interface directly with their switches, as we were an independent telephone company. Effectively we wanted the right to interface with any level of switch that was appropriate.

In a letter to me from Bell dated March 11, 1984, Bell did attach a draft Interconnection Agreement that went part way.

In CRTC Decision 84-10, the Commission concluded that the interconnection of cellular radio systems to the PSTN (Public Switched Telephone Network) was in the public interest. The Commission considered that restrictions on the ability of competitive mobile systems to choose the most cost-effective routing of inter-system traffic were inappropriate. The Commission did not make a ruling on the sharing of toll revenues, and did leave the allocation of phone numbers with the wireline company.

While this was significant progress, the latter was something of a disappointment. Bell proposed providing relatively small blocks of telephone numbers and did not, as we had requested, provide an entire area code for cellular. This would have been far more efficient for mobile systems, as one could then readily identify that it was a mobile phone number by the area code. The result was that mobile phones, with their rapid growth, quickly chewed up existing area codes, necessitating the relatively early introduction of area codes such as 647 in the Toronto area.

However, we would take progress where we could get it.

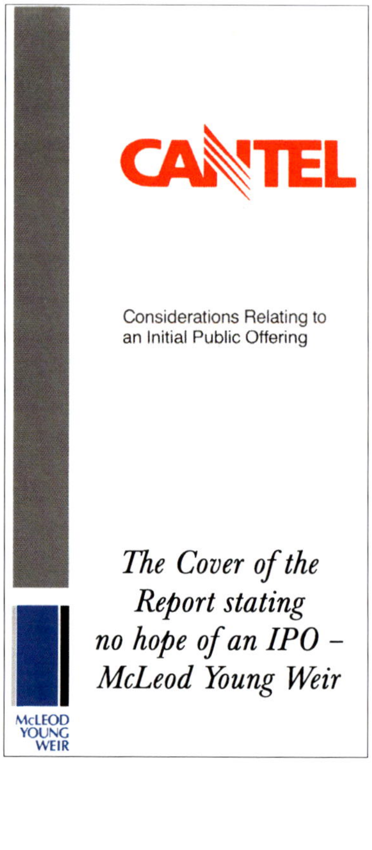

The Cover of the Report stating no hope of an IPO – McLeod Young Weir

Paying For All This

The company was starting to burn through money. The founders had put up $3 million and we had borrowed the rest from the TD, for a total of about $5 million. However, this clearly was not going to even start to finance the actual build.

We were sure we could get vendor financing from whoever our chosen supplier was (NovAtel was, of course, our initial choice). We assumed that an IPO of some description would be possible, but our financial advisor, McLeod Young Weir, told us frankly that given where we were in the first quarter of 1984, an IPO was out of the question.

Further, several attempts at private placements in

Canada were equally unsuccessful. No one seemed to believe the projections, and people seriously questioned whether anyone would pay per minute for a service they receive for a monthly fee on land line phones.

A desperate situation led to a desperate move.

Far Fields Look Greener

With the collapse of our initial attempts to get financing in Canada, we had to look at any alternative.

On April 12, 1984, before Walter had arrived on the scene, the partners, Chuck and I had a luncheon with DOC that included Francis Fox and Vince Hill. We also invited John McLennan, feeling he would be able to give a good outside perspective on the industry.

John McLennan

It was clear that the ever-perceptive Vince Hill realized we were going to have difficulties getting Canadian financing, and I suspect DOC still had some concerns about CNCP being elbowed aside during the licencing process. He suggested we might consider approaching CNCP. Given our concern that CNCP had never demonstrated a real entrepreneurial thrust in this area, we were not warm to the idea. But it did open up the possibility of bringing in another Telco partner.

It was at that rather important lunch, held in the Westin Hotel in Ottawa, that Vince Hill also agreed we could apply for private microwave, as we were a carrier.

Although there was no single ownership direction to CANTEL, it had been agreed that Philippe would take prime responsibility for the strategic planning for CANTEL, Mark would be responsible for the financing, and Ted for the engineering and operational aspects. All of this would be 'coordinated through the Chairman'.

With that in mind, Marc had approached Motorola, who had been the partner with CNCP in the bid but now had no entry point into Canada. Motorola offered substantial (80%) financing for the equipment providing we, of course, bought it from them. This arrangement was no better than we thought we could get from any other supplier. Further, they wanted a 20% equity stake but would put up no money for this.

Marc also approached the American Information Technologies Corp., better known as Ameritech, one of the seven U.S. 'Baby Bells'. Ameritech immediately saw the advantage of having a roaming partner north of the border, as well as the huge potential Canadian market with very attractive licence conditions. Originally Ameritech said they would be willing to invest $30 million for a 30% stake. However, under Canada's foreign ownership laws, an off-shore company could only own 19.9% of a telephone company such as CANTEL.

AMERITECH MOBILE COMMUNICATIONS

Ameritech Mobile Communications

Ted and Marc met with Denny Strigl, Ameritech Mobile's President. Denny would make the ultimate decision about an investment in Canada. It clearly intrigued him. First, because it would be the only foreign investment by a U.S. Baby Bell in the cellular business. Secondly, because the Baby Bells were restricted from doing R&D (this was left to AT&T after the split), they viewed CANTEL as a wonderful place to develop new applications.

We used the argument with DOC that such an arrangement would bring particular advantages to CANTEL, and hence Canada, by importing expertise from the United States. Ameritech had been one of the pioneers in the U.S. mobile telephone industry.

But even this was not to go smoothly. We now need to look again at the technology, which was providing its own problems.

SWITCHING SWITCHES

Less Fit Than A Sixty Year Old Swede

Even before the arrival of Walter, on April 19, 1984, I had written a memo to the partners that NovAtel would simply not have enough features available by July 1, 1985 to allow us to compete with the Telcos, who would be using the Northern Telecom switch. Worse than that, NovAtel was nearly double the price of the lowest bidders. We had advised DOC that, despite our commitment to use NovAtel if at all possible, we had gone out for competitive bids. Ranked lowest to highest in terms of cost these were:

- Harris
- Northern Telecom
- AEL Microtel
- Motorola
- NovAtel

The NovAtel bid was approximately double the Harris bid.

Marc immediately pushed us to switch to Motorola. However, Motorola, while well-known for their telephones, had little experience in large-scale telephone switches. Further, they were anything but a Canadian company and had shown no inclination to make a major investment in Canada.

Motorola

The next logical choice was Northern Telecom. Over the summer, there was considerable pressure from Ottawa to go this route if we could not make a go of it with the yet to be developed NovAtel system.

Northern Telecom

We had visited NovAtel in Calgary and were absolutely shocked to see the state of development. Essentially the switch was a 17" R&D rack with wires hanging everywhere. It was clear there was no way they would have anything ready in time for installation and testing by July 1, 1985.

However, Northern Telecom had its downsides. First, as it was the system to be used by most if not all of the Telcos, it would be very difficult for us to differentiate our system from theirs. Secondly, when Nick pushed Northern Telecom to allow us access to their software to make modifications, they refused. This was a relatively common procedure, and their refusal to show any flexibility made it clear that Northern Telecom was really in Bell's pocket. At this point, Nick made use of his long-time association with Ericsson, a pioneer in the telephone switching business. Ericsson was based in Sweden and was a bit more than a 60-year-old Swede. In fact, they had been in existence for more than 100 years.

Ericsson

We now had a double problem with DOC – if we accepted the Ameritech offer, we would no longer be 100% Canadian and hence really no better than some of the other bidders. Secondly, if we went with Ericsson, we were no longer using a Canadian manufacturer. The beaver was getting bruised!

Walter and Nick saw no alternative. Ericsson had an excellent reputation, mature and tested equipment, and could deliver and install in time for the July 1, 1985 start.

Ted saw this differently. He wrote a stern letter to the other founding shareholders that said: "RTL will veto any proposed change from Northern Telecom for Montreal and Toronto."

Ted went on to say that he felt there was no available time to consider another manufacturer. For once, Ted was wrong. He backed off when he learned that an innovative arrangement had been developed with Ericsson by Walter and Nick. This involved Ericsson supplying the switching equipment to CANTEL, while Ericsson would market NovAtel mobile telephone units internationally. The question of using NovAtel switches beyond Toronto and Montreal was left open.

Further, Ericsson had undertaken to do major research and development in Montreal. At first this seemed a bit shallow but it turned out to be a wonderful arrangement for CANTEL, Ericsson and Canada. The Ericsson R&D facility in Montreal grew to be one of their largest and most effective units, now employing over 2,000 people.

Ericsson Research Lab, Montreal

Back on the financial front, on August 14, 1984 CANTEL's underwriters, McLeod Young Weir and Merrill Lynch, had finally prepared a Private Placement Offering Memorandum. This involved selling about 4.3 million Class B Non-Voting Shares representing 30% of CANTEL's common equity at $9.33 each. This would have raised a total of $40 million.

As noted earlier, despite the initial enthusiasm, the Issue never got off the ground.

DOC had indicated that they had no problem with the issue as long as any American investment through Merrill Lynch, the U.S. agent, was kept to a relatively small amount.

It was when this effort failed that the Ameritech proposal was the only one left on the table.

Now a new challenge arose.

The Massé Mess

Brian Mulroney swept into power in the midst of all this. On September 17, 1984, a new federal government was installed. Francis Fox, who had been both understanding and supportive, was replaced by Marcel Massé. Marcel was from a Quebec riding, and a declared nationalist who was dedicated to enhancing the cultural and economic participation of the province of Quebec. It was Marcel who was going to make the final decision allowing CANTEL to bring in a foreign investor and use a non-Canadian prime supplier.

The Honourable Marcel Massé

Philippe and the Belzberg family were Liberals, but fortunately Ted was a solid Conservative and a personal friend of Brian Mulroney from Ted's days as President of Young Progressive Conservative Association. This helped in at least getting us in front of Marcel, and he eventually signed off on Ameritech's investment. He reiterated the ownership limit of 20%. This would take place by having Ameritech invest $14.8 million and having the founders guarantee that CANTEL would raise a minimum of $25 million in Canadian equity. Ameritech could then invest the remaining $6.2 million, which it put up temporarily in the form of a $6 million debenture. The founders had already put up some $9 million and now had to raise at least $15 million more.

Marcel gave them until December 31, 1985.

On December 14, 1984, Ameritech concluded the deal acquiring 19.9% of CANTEL through a newly created Canadian subsidiary called Pan-Canadian Communications Inc.

Marcel also approved the Ericsson/NovAtel deal, although he did point out in a letter dated October 26, 1984 that CANTEL's plans "represented a significant departure from CANTEL's original commitments."

On January 28, 1985, Marcel issued another stern warning noting, "The purpose of this letter is to raise with you my serious concerns

The NovAtel phone Ericsson agreed to market

regarding the level of industrial benefits to Canada, which appear to be resulting from your current procurement plan."

Ericsson responded magnificently. They clearly saw the Canadian market as being their entry into North America. In a Telex dated March 5, 1985, Ericsson assured the Minister that it would buy at least 5,000 mobile telephones from NovAtel and planned to buy as many as 32,000 during the next couple of years.

They reiterated their commitment to establish a major software development centre employing "at least 30–40 scientists" in Montreal. Finally, they noted that they would endeavour to have NovAtel manufacture radio base stations in Canada for sale throughout North America.

For the moment, the Government reluctantly accepted all this.

SELLING CELLULAR

As far back as September 21, 1983, I had written a memo discussing, amongst other things, the relationship with the RCCs. To quote: "The management group has consistently been concerned about the involvement of the RCCs in the organization. Our preference would be to totally control the sales and service organization ourselves. However, we acknowledge that it might make sense to make use of the RCCs on an agency basis."

Ted wanted to have CANTEL develop its own sales organization because, as usual, he would have preferred total control of sales, and hence the quality.

Walter, assisted by Paul Kavanagh, felt strongly that an RCC agency arrangement would be the only way we could launch over a wide area in the time available. We would otherwise face a huge hiring and space location challenge. It is well to remember that we would not just be looking for storefronts from which to sell the product, but as the early installations would be nearly all 'in vehicle', we needed car installation bays.

Further, Walter felt that cutting the RCCs out of the process would alienate them and simply drive them to Bell or the other Telcos.

Speed to market would be essential and this was Walter's strong point.

Walter often used the expression that when facing tough, established competition, one had to play 'press ball', i.e., really take it to the competition.

The partners agreed, Ted reluctantly.

Coming Alive In '85

The name CANTEL Service Centre (CSC) was selected, and Walter and his team set about immediately recruiting as many of the CSCs as possible to be exclusive agents for the newly proposed CANTEL system.

On Wednesday, October 17, 1984, a total of 34 representatives of the RCCs from Toronto and Montreal arrived at the Sheraton Centre in Montreal for the formal sales launch under the banner 'Coming Alive in '85'.

It was a show not quickly forgotten.

As the group entered the conference room they were greeted by upbeat classical music, fully bilingual posters, themed buttons and substantial presentation materials.

The meeting commenced with a 12 projector audiovisual opener that portrayed the history of communications from Adam and Eve on.

Cover of the CANTEL Service Centre material, October 18, 1984

I had hosted an opening dinner the evening before at Le Fadeau, with an eight-course gourmet French meal. I spoke at the dinner expressing my enthusiasm about the challenge ahead.

It was now Walter's turn to chair the meeting the next day. He was superb.

He was followed by presentations from Paul Kavanagh, Vice President Marketing, Joe Church on Corporate Development, Nick Kauser on Engineering, Mark Ferland on the distribution network, Kathy McLaughlin who had joined as Manager, Marketing Communications, and David Parkes who had been added to the team as Vice President and General Manager, Central Region.

Following the afternoon presentations, there was another dynamic a/v emphasizing the challenge ahead for the new CANTEL team and displaying photographs of each of the participants, which had been taken the evening before. The CSCs were blown away.

David Parkes, Vice President & General Manager Central Region

Kathy McLaughlin, Manager, Marketing & Communications

Comments flowed in during and after the presentations on the thoroughness, professionalism and 'sheer volume' of the information imparted. The general comment was "I had no idea so much work had already been done!"

The CANTEL Service Centre concept had been launched in a style that would become the signature of CANTEL. Whatever concerns there were about our lack of telecom experience, financial strength or our ability to succeed had been swept away.

The same energy and spirit that had carried us through the licencing process was now propelling us toward a launch that no one thought we could pull off.

The CSC Presenation Room had a Paris theme

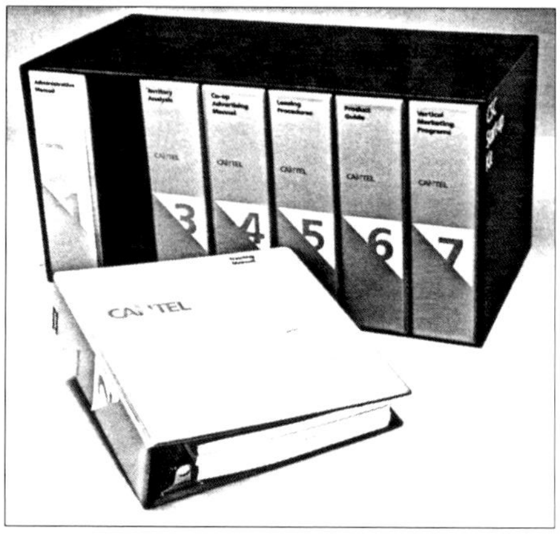

The CANTEL start-up kit for the CSC's October, 1984

Rolling Out The Rollout

Innovation was to be the name of the game. We now had a reliable supplier in the form of Ericsson. We had a pumped up management team with new employees coming on almost daily.

We had relocated our offices from 20 Queen Street to 40 Eglinton Avenue East in a new avant-garde building. We had the top several floors to accommodate the new group.

It made an interesting contrast with the rather dowdy looking Bell Switching Centre a couple of buildings to the east.

As Nick, Roger and others had quickly realized, we would have to set up many cell sites in a very short space of time. Nick's solution was to create pre-fabricated sites that could contain the cell site equipment, HVAC and backup power, as we felt it would be essential to have non-interruptible power if the cell sites were going to be fully reliable. As no suitable fabrication was available for this, Nick suggested the use of sea-going containers to house the equipment. These were strong, weatherproof and readily available. They only needed doors cut into them for the equipment. We contracted with a Montreal firm to equip these, and 18 were delivered on incredibly short notice at a cost of $17,750 each.

This was only one example of the momentum of the newly created team.

A Pre-Fab Cell site being delivered

Our new location at 40 Eglinton Avenue East as it looks today

The Bell Building at 72 Eglinton Avenue East also as it now looks

Snatching Defeat from the Jaws of Victory

With everything going right for once, we then proceeded to mess it up. The relationship between Walter and the three partners was, as forecast, stormy from the start. Caroline outlined the unbelievable meeting that had taken place at the Royal Canadian Yacht Club not long after Walter's arrival. Walter had outlined his intended organization of the company, which included a description of his matrix management approach.

Ted commented that, "This sounds like Communism to me." Philippe noted that it seemed like motherhood to him.

Marc added that he did not understand the problem. His view was that management was simply telling people what to do, and if they did not do what you wanted, you fired them and got somebody who would.

The management group was in the room at the time and was horrified at this confrontation. They were asked to leave the meeting, leaving just Walter, me and the partners to try to straighten this out.

Eventually I went outside to round up the group and assured them everything was fine and this was just a normal disagreement amongst entrepreneurs. However, the rest of the day went badly and the management group left, wondering what it was they had joined.

"It was a horror. It was an absolute horror," recalled Walter Steel.

It went from bad to worse. As noted earlier, Ted in particular took exception to the move to Ericsson, and Marc was disappointed because he was pushing Motorola.

Ted disagreed with the idea of CANTEL Service Centres and would eventually win that argument, but not for another five, or so, years.

Walter felt the partners were going over his head to cut a deal with Ameritech, although he did not disagree with the approach. In fact, Walter and his group were elbowed out of the negotiations on the final deal. This went to Oliver Bush, who was Marc Belzberg's proxy.

Walter interpreted all this as being a clear message that the managers were there solely to do as they were told. He concluded that the owners didn't trust each other much, let alone the hired help.

When Denny Strigl attended his first Board meeting following Ameritech's investment, he could hardly believe the wrangling that took place at the Board. Board meetings sometimes dragged on for two days, with the partners barely able to agree on anything.

On February 14, 1985, Walter was asked to meet with Ted at his home at 4 Frybrook. Ted, who viewed Walter as a major part of CANTEL's perceived problems, suggested that he should quit. Walter told him to go fly a kite.

However, that was the end of the relationship. I had always got along quite well with Walter and respected his determined approach to get the job done. I was still Chair; Harold Nickerson, who had recently taken over as President of Télémédia, had been appointed Chair of the newly revamped Executive Committee.

It was up to Harold and me to finalize the departure of Walter. The three of us met in the offices of Tory, Tory, DesLauriers & Binnington, and by that time Walter saw there was no point in persevering. He thought of suing the owners, who had no intention of honouring his stock options, but thought better of it. The meeting at Tory's was actually quite short.

As Caroline points out in her book, the firing "went down in CANTEL's lore as the St. Valentine's Day massacre."

Demonstrating a hand held Motorola outside our new office

Walter and I parted on as good terms as could be expected under the circumstances. He seemed to have developed some respect for me as a calm presence in a rough sea and as someone who tried to be fair. I, in turn, felt he had done a very good job in the short time he had available, and under circumstances that would have driven most to drink.

We were now only four months from launch and were again without a President & CEO.

John McLennan, who was on our Board at the time, agreed to replace Walter as interim President long enough to see CANTEL through its July 1 launch. John did a very fine job of holding everything together and restoring the momentum that was nearly lost with the firing of Walter.

Amazingly enough, the team held together. This was certainly helped by John's personable and yet businesslike approach. But this was also aided by the adrenalin rush of completing the job that no one expected us to be able to complete.

We Have Lift-off

On July 1, 1985, we launched as promised in Toronto, with a corridor from Hamilton to Oshawa and in Montreal.

Ted and I had arranged with Art Eggleton, Mayor of Toronto, to

Art Eggleton, Mayor of Toronto, with George making the first cellular call, July 1st, 1985

make the first phone call on cellular from Toronto to Montreal. Mayor Drapeau of Montreal was at the time in the Ramses II Exhibit on the Expo site to participate by receiving the historic call. I noted to the press afterwards that this was likely the first cellular call ever to an Egyptian tomb!

Mayor Jean Drapeau of Montreal receiving the call – Marc Ferland and Joe Church in the background

The result was a clear victory for CANTEL. The Canadian Communications Network Letter of August 26, 1985 estimated that in each of Toronto and Montreal, CANTEL had substantially outsold Bell.

Bob Rowland, President, Bell Cellular, and Paul Kavanagh each noted that the figures quoted were far too low, but there seemed little doubt that we had jumped ahead from day one.

As I pointed out in the CANTEL Communiqué, a regular publication for employees, "In the cellular industry – as in any two-team league – the measure of success is simple: you're either #1 or you're last."

It turns out that we had not only outsold Bell, but independent analysis showed that we had also out-engineered them. It was the latter that infuriated them more than anything else.

The next big challenge was launching in Ottawa and Hull. In this case Bell Cellular beat us to the punch by about two weeks, but we were soon up and running in the DOC's home territory.

Once launched, however, John had announced that he wanted to step aside. Again, CANTEL was in search mode for a new President.

The Flip/Flop

Denny Strigl had taken over as Chair of the Executive Committee and led the search for a new President. After some discussion, as Caroline noted, "The owners turned to the avuncular and mild-mannered George Fierheller, whom they felt they could trust even though he was clearly a Rogers man."

I had to look up the word 'avuncular', which turns out to mean "an older man, benevolent and friendly." This was likely not a bad definition of me.

I did everything I could to distance myself from Rogers. I resigned from all Rogers positions, including their Board, and even sold any shares that I had.

The downside was that I now had to move from Vancouver to Toronto, as the former location would be impossible for running an organization the size that CANTEL would likely be.

John was to be the new Chair, i.e., we flipped positions, but he was not available on September 1, 1985 when I took over as President & CEO. It would have been logical for him as Chairman to have welcomed me as the new President. Instead, I had to welcome myself, and did exactly this at our 40 Eglinton Avenue East office at a meeting of all Toronto employees. I shook my right hand with my left hand and congratulated me on the new position.

I was now back where I started as President of the company from which I had tried to extricate myself.

The new/old President

PART THREE

Back in the Cellular Saddle

The New Cellular Cycle

Just as happened after the win of the licences some 18 months earlier, the post-July 1 period was a bit of a letdown. It is hard to imagine the intensity that went into getting the new cellular system operational by the July 1 deadline.

In Toronto, we took the staff and the CSCs in the area on a 'thank-you' cruise on the Mariposa Belle – a ship that sails in the Toronto Harbour. However, we did not have long to celebrate as we had committed to roll out the system in the Ottawa area in September. As the recycled President and CEO, the cross-country rollout was to be my major commitment for the next couple of years.

Even before Ottawa, we had to address a growing 'summer' demand to service cottage country. This was compounded by Ted wanting service to his island on Lake Joseph, and so we started the build north of Toronto along Highway 400. It turned out that Jean de Grandpré of Bell was making similar demands to Bell Cellular for service north of Montreal, and to match their build, we had to expand in this area as well.

I dress for the occasion

Finding locations for cell sites was not easy. For example, as we endeavoured to complete our service between Toronto and Montreal along Highway 401, there was a gap around Colborne. Although we had located several suitable sites, the local Council decided they did not want an unsightly cellular antenna in their area. I decided the best approach was to appear before the Council to explain that they did not want to be responsible for the 'Colborne hole', in the coverage as this would bring all kinds of unwanted criticism. I found out when the Council was meeting and asked to attend personally. I correctly guessed that the dress code would be informal and somehow located an old plaid shirt. It turned out to be a good guess and I actually found the local municipal officials quite appreciative of the fact that I tried to be one of them.

I explained that while there was some flexibility in where we put towers, the optimum spacing and a location close to the highway were essential. They ended up helping us with some of the farmers in the area, a number of whom were quite happy to have the cell site as we, of course, paid rental for the land used. The problem was solved, but this was typical of the painstaking process of finding cell site locations. Nick Kauser and I used to joke that we had never in our lives looked at the roofs of condominiums and now found that we were examining them everywhere we searched for possible locations.

We located our first Toronto switch at the foot of the CN Tower. This was not because we could use the Tower as a cell site, as it was far too tall – remember that the principle of cellular was to keep the signal from going too far or it would interfere with adjacent cells. However, the CN Tower was an ideal location for microwave shots to Oshawa, Hamilton and elsewhere in Ontario, and these were used to link the cell sites to the switch.

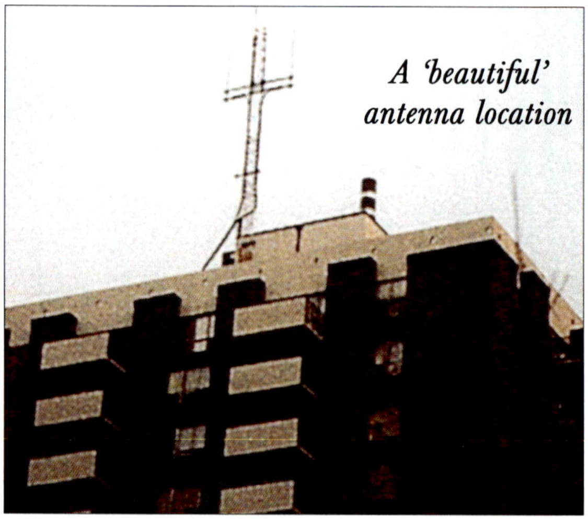

A 'beautiful' antenna location

On September 5, 1985, I announced that in addition to being in commercial operation in Toronto, Montreal, Ottawa, Hamilton and Oshawa, we now had a network of 29 cells. We also had 156 employees, not counting those involved in our 19 CANTEL Service Centres.

We had been active in developing exclusive promotions with such organizations as Esso, American Express and Canadian Tire.

Shortly after the launch in Ottawa, we commenced service in Quebec City. We arranged to celebrate the announcement of the service by my dropping the puck at the opening of a Quebec Nordique hockey game. I presented the captain of the team with a complimentary cellular phone using my very limited French as best I could.

The next big launch had to be in British Columbia, given the involvement of the

The microwave dishes on the CN Tower

Belzberg's. Sam and Fran had held a wonderful launch lunch. Fortunately I knew the principals of B.C. Telephone very well from my recent days in Vancouver and they were quite cooperative in negotiating an interconnect agreement.

Service launched in B.C. on January 10, 1986. We had hired Brian Josling as Vice President & General Manager, Western Region, and he was instrumental in helping us move forward with the next phase of the launch, which was into Alberta.

Despite the involvement of the Belzberg interests in that province, it was not an easy sell. Marc was concerned that Alberta Government Telephone (AGT) had a well-established monopoly, particularly in the oilfields, with their long-established mobile service. They had about 25,000 customers at the time. Marc felt it was a waste of money trying to compete in the province.

I persisted, however, constantly pushing the concept that we had to make the best use possible of our national telephone licence. I made it clear that it was my intent to roll out across the country as rapidly as possible.

The rest of the Board agreed that this was too good an opportunity to miss, and we started operations in Edmonton on September 1, 1986.

I met with Ralph Klein, then Mayor of Calgary, and convinced him to support our opening in that city and we added service there in January 1987.

The launch in Alberta was slightly clouded by our relationship with NovAtel. Although we were still promoting the NovAtel phone, it had now become clear that they would never be successful in the switch business. CANTEL would be a dedicated

The view from our Vancouver switch location

Ericsson user throughout its entire network. I met with the late Bob Blair, who was by then Chair of NovAtel, just to reiterate that we were doing our best to promote an Alberta-based company through the support of their phones but that was it.

It was a challenge dealing with any of the provincial telephone monopolies. In Manitoba, the Legislature had actually passed a resolution preventing CANTEL from opening up in the province. I met with Gary Filmon, the PC Premier at the time, and he was actually quite open to the idea of competition. He had already privatized some Manitoba public corporations and he was very helpful in getting the Government's position reversed. I also met with Gary Doer, who was then Leader of the Opposition and later Premier. He also turned out to be supportive as I think he realized that the choice was either allowing CANTEL into the province or not having cellular at all.

We opened the system in Winnipeg in May, 1988.

Saskatchewan was even more difficult. I had met with Grant Devine, who suggested I meet with the Minister of Communications. I got nowhere with him, and his final comment was that this would be a decision by the Minister of Finance. "But you are also the Minister of Finance," I noted. "Now you get it," he replied!

However, the DOC ruling proved powerful enough that the government ultimately let us open in Saskatchewan in August 1989.

By July 1987, CANTEL had become the first cellular company to provide continuous coverage of the 1,300-kilometre corridor from Windsor, Ontario to Quebec City. By November of that year we were also operating in Nova Scotia, and had a national

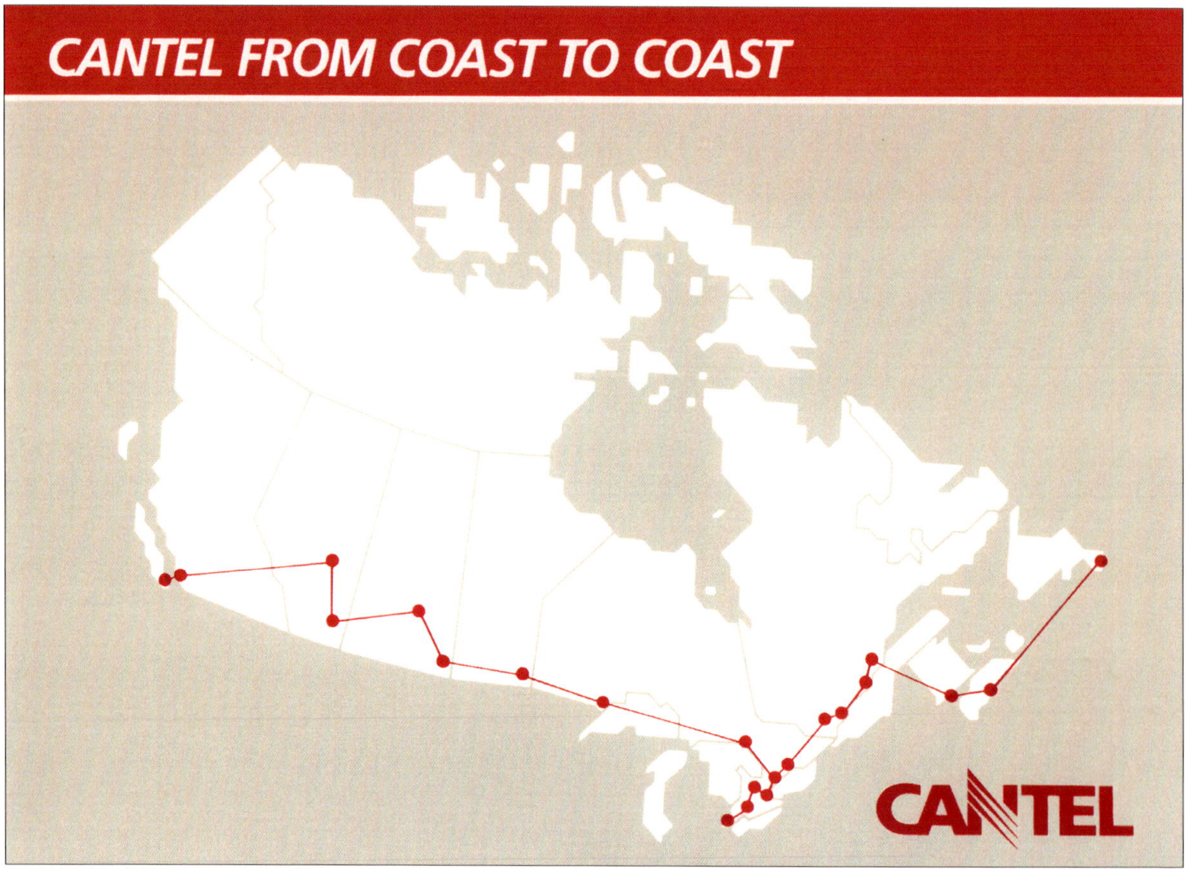

network of 38 CANTEL Service Centres. In October of 1988, we had reached the 100,000 customer level. I remember that we selected Dr. Julia Weste, a Halifax veterinarian, as our 100,000th customer. It seems incredible that now Rogers Wireless adds more than 100,000 new customers each quarter.

I continued with the remaining provinces, having met with Frank McKenna who listened politely to my arguments for allowing CANTEL to start in New Brunswick. He then pointed out the window to the head office of NBTel and said, "When you can offer as many jobs in the province as NBTel, we would pleased to have you here." We opened in any case and completed our across the country coverage in Newfoundland in June 1990.

In the summer of 1989, I realized my final ambition when Ted Rogers announced the Coast to Coast TransCanada Cellular Corridor of 7,500 kilometres – the longest continuous cellular telephone network in the world. This was a $600 million investment by RCI, who had already invested over $1 billion in CANTEL.

During those few years I think I made more speeches than Barack Obama during his campaign to become President. In trying to introduce this new field of Wireless Communications to as broad an audience as possible, I must have spoken to every Lions Club in the country. I particularly remember one in Halifax, where I had tried to liven up the topic by talking about the origins of communications and why this was such a revolutionary concept. At the end, one elderly Lions Club member came up to me and said, "You made a dull topic almost interesting." I think this was a compliment.

At one presentation, John McLennan was introducing me and referred to me as the "grandfather of cellular in Canada." I was a little taken aback by the grandfather reference, thinking that 'father' would have been better, but could hardly object as I was already in my mid-50s by that time.

The whole rollout was a challenge but, in retrospect, it was also a lot of fun. This photo shows me at a cutting-the-cord ceremony in Fredericton.

Let Me Say This About That

The Fredericton launch

CALLING CANTEL

I have been emphasizing the rollout across the country, but should backtrack to look at what CANTEL was actually selling. I had indicated I would try to outline what I thought differentiated us from the Telcos and certainly the product development and promotion were big factors.

To begin with, the new CANTEL logo, which we had started to use in December 1984, was upbeat and lively relative to the more staid logos of the Telcos. It worked well and when we launched, we had CANTEL spelled in giant, illuminated letters at the top of the CN Tower. I had chosen a rather impish symbol of a small car with a cellular aerial, which Ted ultimately objected to as he thought it was a bit frivolous when we were trying to compete against the Telco giants. I still feel, however, that it was the right David/Goliath approach.

Cover of Employee Handbook showing the original symbol

The corporate colour was a brilliant, slightly blue-toned red. We used this until the time of the amalgamation of all logos with Rogers. Prior to this, the Rogers Cable Company had used a very low-key colour referred to, for some reason, as Grabber Orange. The whole Rogers Corporation ultimately adopted the CANTEL red, but slightly altered the tone, to make it a bit more of an orange red (perhaps to provide some linkage to the past) but it was not as vibrant as our original choice, in my opinion.

But, beyond logos, the 'Phone Company For People On The Move' was being particularly innovative in its development of features. As early as June 1985, we were offering exclusives such as Call Forwarding, Call Waiting, Conference Calling and even a telephone directory listing.

The late Maureen Ladly, who we had hired from Bell to head our Customer Service, developed a detailed customer billing, which we thought would be a competitive advantage. She pointed out that this was such a valuable addition for our business customers that we should

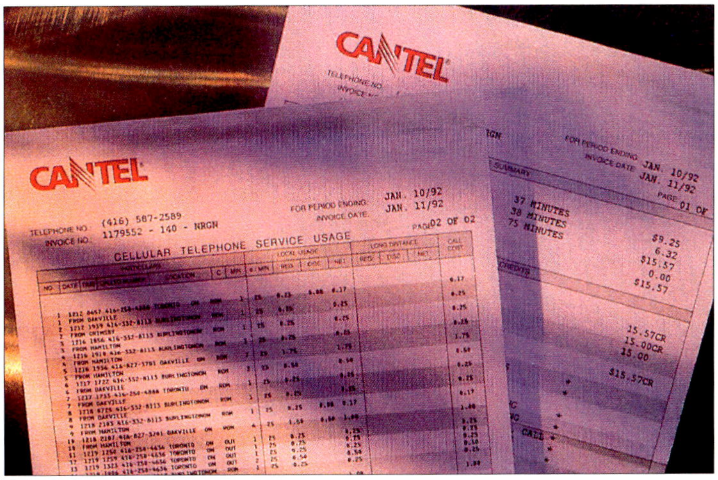

CANTEL detailed invoices

actually charge for it. It seemed audacious to me to charge customers for their billing but Maureen was absolutely right. Not long after announcing this service, we had more than 10,000 business customers who were actually paying for the details of the phone usage on their invoices.

Roaming had become a very big factor, and roaming into the United States through our connection with Ameritech and later, McCaw, was very important. However, as early as October 1986, we actually established the first international roaming agreement with Hong Kong.

Under Paul Kavanagh's direction, we launched one of the first customer retention schemes with the CANTEL Club. This was a highly successful program that allowed users to accumulate points and cash them in for either free minutes or merchandise. Again, Ted, who was known to micromanage, felt this was an unnecessary expense and, after acquiring control of 100% of CANTEL Inc. in February of 1989, discontinued the program. It was at that time that the company was renamed Rogers CANTEL Mobile Communications Incorporated and the stand-alone name CANTEL that had served us well up to that point, was diminished.

It might sound as though CANTEL was doing everything right at this point and the Telcos were just scrambling to keep up. However, CANTEL was doing its own scrambling.

You may recall that the early projections made by each of CANTEL and Bell were conservative, with Bell being considerably more conservative than we were. We both, however, turned out to be very wrong. At the date we had projected having 100,000 customers, we had exceeded 260,000. The network was falling behind and it took a huge push to develop enough new cell sites to meet this ever expanding demand.

The customers, particularly in Toronto, were complaining. I set up the 'Office of the President' to try to handle as many complaint calls as I could, many of them coming from my YPO friends. The Rogers Cable Company, which had contacts in many condos because of their services to those buildings, was particularly helpful in locating new cell sites for us. We skated through this crisis but it was not an easy period.

Cell Coverage Maps – matching the demand was a constant challenge

Momentum Through Motivation

When you surround yourself with bright, young, energetic people, it is not a big challenge keeping them motivated. However, if our initial lead over the Telcos resulted from Nick's engineering, a solid marketing program with the CSCs and the string of innovative product offerings, what was going to sustain this momentum was the morale of the group.

I felt that one of the main jobs of the President was to create an atmosphere in which new ideas were expected and employee recognition was given at every opportunity.

In those years, we put out an annual booklet with everyone's picture, and encouraged each local group to provide anecdotes about what was going on in their area.

As I said in the President's Message to the 1988 Yearbook, "CANTEL is more than a collection of switches, cell sites and Service Centres. What makes this company come alive is our spirit.

"Our people continually inspire each other to greater heights of creativity and accomplishment.

"You can't place a dollar value on what this means to a company but it is what makes us so formidable."

In the 1989 Yearbook my message began by noting, "You can lead an employee to a desk but you can't make him think!

"In fact, unless an organization is a fun place to work, you can't make people do much of anything. CANTEL's style has always been to provide a challenging, interesting, and lively environment that encourages people to do their best."

The cover of the 1989 Yearbook even had the first verse of our company song:

"Person to person, day or night, CANTEL is doing it, doing it right."

But this spirit did not just happen. It had to be nurtured.

I continued a tradition I had started at SDL and in some prior leadership positions of regularly communicating with all of the staff. When the company was small, I did this through monthly Fireside Chats. It was reasonably easy to get the staff together in Toronto, Montreal and Vancouver. Eventually, we just had too many offices to make this practical and I took to broadcasting these, but making sure there was lots of time for interactive comments from the field.

Naturally, not everything went smoothly. For example, at our new location at 40 Eglinton Avenue East, we were growing so rapidly that we needed additional space. The landlord agreed to let us use part of an additional floor but without any improvements, i.e., no ceiling tiles, carpets or other amenities. The price was right but the employees who were working there lightheartedly referred to it as The Dead Zone.

I wanted to encourage the staff to bring up any problem such as this or suggestions for improvement, and started to hold Brown Bag sessions by announcing I would have a sandwich in the cafeteria on certain days and be available to anyone who wanted to talk about anything at all. This group of young tigers did not let the opportunity pass. I thought they might be somewhat reluctant to talk to the President, but this was certainly not the case.

1988 Employee Yearbook (me in the balloon!) – 5 years after winning the licences

1989 Yearbook Cover – complete with the CANTEL song!

I then found an even more effective venue was the Brown Bottle approach. On Friday afternoons, I would join a group of regulars at the Chick 'n' Deli around the corner on Mount Pleasant. The chicken wings were hot, the beer cold and the feedback I got was invaluable.

The Open Door policy was always there, but I really felt you had to go to the staff rather than waiting for them to come to you.

By 1998, our revenue had grown to more than $100 million, with the number of CANTEL Service Centres close to 40. Our subscriber base had doubled that year from 45,000 to 95,000, and our coverage was now in 35 centres in six provinces.

We had to relocate our executive offices from 40 Eglinton Avenue East to 10 York Mills Road. I had chosen this location because it was right on the subway and had good access to Highway 401 and hence the airport – the staff hated it.

Yonge and Eglinton had been a lively area with lots of restaurants and shopping. York Mills was somewhat isolated, with nothing but a mediocre food court. However, it did provide room for growth and we soon occupied most of what was a very large building.

I should point out that we had undertaken to DOC to locate our legal Head Office in Quebec. We had selected a location at 6315 Côte de Liesse, St-Laurent. This was actually a good location, being reasonably close to the airport and it was our initial switch location in the Montreal area.

10 York Mills

There was an office set aside for the President but, in reality, that ended up being a lunchroom most of the time, as Walter, in particular, did not even try to run the company from there. In fact, Joe Church, who we had hired from Bell, was the resident executive in Quebec except, of course, for Mark Ferland who was Regional Vice President.

Customer morale was just as important as employee morale. After we opened in Vancouver, we commissioned a Toronto

The Customer 'Thank you' Picture - Gérard Paraghamian

artist, Gérard Paraghamian, to do a composite rendering that combined the skylines of Montreal, Toronto and Vancouver, with a couple of sailboats in a waterfront scene. The only reference to CANTEL was a very small company name on the front of a sailboat. These renderings were given to our major customers as thank-you's and appeared on a number of office walls.

I might add that there was another group that required constant hand-holding. The downside to having independent agents was that they were independent minded. Ted had correctly forecast that this would be a problem, but neither Walter nor I, nor the marketing executives at the time, could figure out a better way to launch in the time we had. The stories of the sometimes stormy relationship would fill a book. I will give you only one example.

One of our CSCs based in Montreal was Abraham Finkel who owned an agency called Celluland. He was effective enough as an agent, but was constantly on the verge of suing CANTEL for one indiscretion or another.

Finally, the sales group came to me and said, "You straighten it out." They suggested I invite Abraham to Toronto for a meeting.

I took a different approach. I called Abraham and told him I would like to see his operation in Montreal and, in particular, meet his wife and family who were involved in the agency. He seemed pleasantly surprised that the President would take the time to do this. I toured their facility, took them out to lunch and did my best to address some of the problems, which then seemed to largely melt away.

Abraham, as it turns out, had been a member of Mossad, the Israeli Security Police, and he was one tough cookie. I was told that he kept a handgun in his glove compartment but never verified this.

Abraham continued to be a challenge to CANTEL for many years thereafter and, on numerous occasions, I was called in to calm the situation down – not ultimately successfully, as he ended up suing us anyway.

Many of our CSCs were, however, very well run and productive. They ranged from Jimmy Pattison in Vancouver to Charles Sirois in Quebec. Charles eventually sold his paging/agency business to Bell and this formed the start of his unique association with that organization. To Charles' credit, however, he loyally turned over all his CANTEL clients to us and made no attempt to convert them to his new partner.

Charles Sirois, President, Télésystem, Chair, CIBC

In summary then, the period from the start-up in 1985 through 1990 was one of concentrating on building the network across the country, hiring and motivating staff in ever-increasing numbers and trying to stay ahead of customer needs.

By 1991, CANTEL had 2,200 employees and more than 350,000 subscribers. We were handling close to three million calls a day.

CANTEL's success had been acknowledged as early as 1987 when Bob de Cotret, Minister of Science and Technology, presented CANTEL with the 1987 Canada Award for Business Excellence in recognition of our introduction of cellular telephone service to Canada.

We were well on our way. But, to where?

An early handheld phone – still not really pocket size

THE NEXT NETWORK

Part of the foresight of Nick Kauser, Bob Berner, Roger Keay and other members of our original engineering group, was to recognize that while Cellular had started as an in-car service, the future was clearly in hand-held units. They had, to the extent possible, designed the network with this in mind. This was not an easy task.

As noted earlier, with only the low power of a hand-held unit to transmit to a cell site, the cell sites needed to be very close together. CANTEL pioneered the mini-cell for use in a building, e.g., where a small cell could be used to service a single floor or an underground parking location.

So that cell sites could be placed as close together as possible to maximize the capacity, as well as being close to the hand-held user, the orientation of the antennae could be 'down tilted' so the signal would not go too far.

Signal strength was always a problem, and one of the challenges with in-the-car units was that people did not like to have the spiral aerial standing vertically from their rear window. I suppose it did not look sufficiently streamlined. They had a bad habit, therefore, of tilting the aerial backwards without realizing that the purpose of the spirals in the aerial was to increase the strength of the signal and if you tilted it backwards, the signal tended to radiate into the air or into the ground. Drivers did not really appreciate CANTEL employees suggesting they would improve their reception if they put the aerial back upright as it had been designed to be placed.

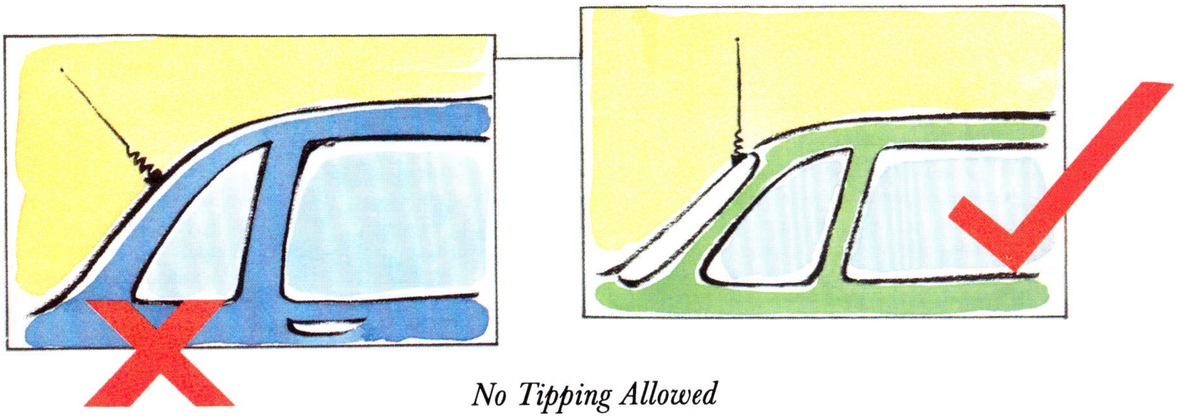

No Tipping Allowed

In any case, while we likely overspent on the network, the payoff of better reception for portables was clear.

The bulky DynaTAC produced by Motorola, and similar phones from NovAtel and others, were hardly user friendly. They were really briefcase, not pocket, phones.

At an Annual General Meeting of CANTEL in 1987, I made the prediction that phones would shortly be of a size that could fit quite easily into a suit pocket or a purse.

I then proceeded to pull out of my pocket the newly miniaturized Motorola MicroTAC. This was a just-announced fliptop product that truly revolutionized the cellular industry. There were oohs and aahs from the audience.

The MicroTAC flip phone with retractable antenna

While this prediction was right on, my forecast at the next Annual General Meeting in 1988 turned out to be completely off base. At that meeting, I pulled a cheap plastic felt-tip pen from my pocket and said this would be size of cellular phones in the future. I noted that the length of the pen was about correct for a portable cellular antenna. It would be quite easy to have a mini microphone in one end and a tiny speaker at the other end. Someone in the audience asked how you would dial it. I had the ready answer that it would be voice activated.

The cellular field never went anywhere near this. As we all know today, what people wanted was more functionality – customized ring tones, mini keyboards, full motion video, built-in cameras and all the other features we now see. While the phones did indeed get smaller, they never approached my famously predicted 'pen phone'.

Oh well, you cannot get them all right!

However, the advent of the hand-held phone did mean that phones could now be sold through outlets that did not have to have installation bays. This in turn would have an impact on our CSC network, as it opened up the possibility of marketing the phones in storefronts.

By March, 1990 CANTEL had created its own retail division by establishing Rogers CANTEL Service Centres. It got its start by acquiring 18 of our franchised CANTEL Service Centres from Glenayre Communications Limited and renaming these as Rogers CANTEL Service Centres. Eventually Ted would get his wish, and more and more of our sales would be controlled internally.

Safety First

The advent of the hand-held cellular phone brought with it some safety concerns. As President of Canada's largest cellular telephone company, I played an active role in the Cellular Telephone Industry Association (CTIA) that was based in Washington. Tom Wheeler, the President of the organization, asked if I would Chair a new committee on cellular safety.

After examining exposures I felt the industry had, I outlined three:

- the safety hazard of driving with a phone in one hand
- privacy issues
- electromagnetic radiation

CANTEL had launched a Hands On! For Safety campaign in 1988 emphasizing, amongst other things, that cellular users should utilize a hands-free speaker system, which we heavily promoted.

We also developed the first network-provided voice activation system to allow people to dial by simply speaking the number or even the name of specific recipient, e.g., secretary, spouse, etc.

Despite these efforts, this remains a contentious issue to this day.

I recall at one point the North York City Council, when North York was not yet a part of Toronto, considering a resolution to ban cellular phones in cars. I appeared before the Council armed with all sorts of statistics that insurance companies did not feel they had to charge more for cars with cellular phones, there were no demonstrated statistics that cellular users were more dangerous than those turning on a radio or inserting a tape, and so on. One of the councillors commented it was likely no more dangerous than drinking a cup of coffee coming into work in the morning, which she did regularly. Another councillor said, "Yeah, I use my electric razor in the car."

The motion was defeated but it did indicate that this was going to be an ongoing safety problem.

The second problem I had predicted had to do with privacy. The analogue signals of the original cellular systems were not encoded in any way. It did not take people long to realize that the frequencies that had been provided by the FCC in the United States and the DOC in Canada were the same frequencies

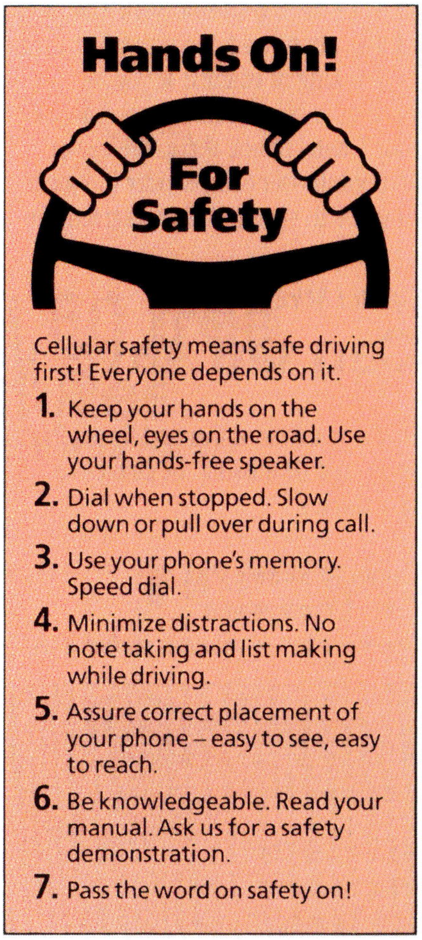

CANTEL Safety Reminders

that had originally been allocated, but rarely used, for UHF television. In fact, if you had one of the old TV sets, you could dial one of the stations and pick up a cellular conversation. As the conversations were handed off from cell site to cell site, you would have to relocate the signal, which really made eavesdropping rather impractical. But, in a pre-digital era, this was a problem.

Entrepreneurs developed cellular listening devices that would scan the frequencies so that one could follow a telephone conversation quite easily. Digital cellular largely solved this problem but, for the time being, it was a challenge.

Neither of these, however, provided the problem of my third predicted challenge – that of a possible radiation hazard from holding a transmitter up to your ear. This, of course, had not been a problem when the transmitter was located a long way from the individual, as was the case in a car, but with the advent of hand-held cellular phones, I felt this was an issue that was going to come up.

It was within weeks of my making these three predictions that the Larry King Live show put on a program where someone claimed they had developed brain cancer as a result of overuse of hand-held cellular phone. In typical Larry King fashion, the show was long on rhetoric and short on facts. However, the CTIA had to scramble to meet the challenge.

Wisely, the CTIA established a Blue Ribbon Panel of international experts to study the potential hazard. They were quickly able to determine there had been no increase, at all, in localized brain cancers even amongst heavy hand-held cellular users. They conducted numerous tests to indicate that the 0.6-watt level was well within any safety limits.

However, the problem remains a challenge for the industry, particularly in the litigious U.S.

What I had assumed would be a relatively minor industry role had now become a major headache (not caused by cellular radiation!).

Cellular etiquette

Dinner etiquette, dating etiquette, even business etiquette, are all part of our lives. So why not cellular phone etiquette? We think it's an idea whose time has come. Here are a few questions posed to our arbiter of etiquette — who, incidentally, invites your questions to be included in future issues of *Connections*.

Coming in December

Q: Is it appropriate to use your cellular phone while dining in a restaurant?
A: Maximum discretion here. If it's absolutely necessary to use your phone, please *lower your voice and keep it short.* Inform your caller of your location. Avoid eye contact with fellow diners! A better bet may be to let Mobile Message take any calls. If not, please turn the ringer volume down.

Q: What do I say to a caller who comments on my handsfree speakerphone, implying that he'd really prefer I talk to him on the handset?
A: Handsfree speakerphones are improving in quality but there is still an audible difference to most ears. The ambient noise in your car also plays a role. But remember, hands on for safety! If you're driving, go hands-free. Pre-empt any comments by informing your caller up front. If it's a really important call and it's safe to do so, you may want to pull over and use your handset.

Q: Is it good manners to ask my passenger to place a call while I'm driving?
A: Safety is never bad etiquette! It's bad manners to distress your passenger by fumbling with the dialer while maneuvering your car through traffic. Ask for help or wait until you reach a stop before you dial.

Cellular Tips

Every Speck of Spectrum

Just as I had pushed for making the maximum possible use of our national licence, even when it involved putting up cell sites in marginal areas so that we could claim close to national coverage, I also felt we should go after every bit of spectrum that became available. I certainly did not need to push Ted, who understood that spectrum was the electronic highway on which all future applications would run. This thinking led CANTEL to enter the paging business with Canada's first 900 MHz paging network on July 1, 1989. This was an advanced paging system, with voice, tone, numeric and alphanumeric paging capabilities.

You might wonder why there was still a role for a pager when one could use a cellular phone to receive the message at once. However, when you think of it, a contractor installing a roof would not likely want to receive an actual phone call. He would be better off just knowing that someone was trying to get in touch with him and having the number available to call back when he was next on the ground.

The Telcos already offered paging services and CANTEL felt that a low-cost entry point to wireless communications would have a role for some time. It proved to be a worthwhile investment and, by October 1989, we had over 9,000 customers.

In line with the idea that we should get all of the spectrum we could, CANTEL even bid on providing air-to-ground service for aircraft.

My belief was that ultimately everyone would have a single device that could operate anywhere in the world, and that device would automatically pick the least-cost alternative for completing the communication. For example, if the device detected it was close to a cellular system, it would utilize that. If it was out of range of cellular service, it would utilize a satellite-based service.

CANTEL Paging – reminds me of current Blackberry use

The satellite service was indeed soon proposed, with a system developed by Motorola called Iridium. The concept was to put 66 low earth orbit (LEO) satellites in place so that a ground-based hand-held unit would never be out of sight of a satellite anywhere on earth. I tried to get CANTEL interested in participating in this service but had to admit that, after careful analysis, we just could not see the business case. We could not forecast that there would be enough users who would be out of reach of a cellular system and would be willing to pay $10 plus/minute for the service. This analysis turned out to be correct but the principle was certainly appealing. I was on the Board of Telesat

Mobile Inc. and was perhaps carried away with the potential of satellite at that early stage.

However, CANTEL had the opportunity to pioneer another type of service. Long before text messaging became popular, it was obvious that digital data communications would have a big role. This could be for anything from taxi dispatching to contact with a network of trucks or even remotely monitoring an oil pipeline.

The difficulty of using a cellular network for this type of communication is that it would rapidly chew up the available spectrum. We all knew the use of analogue spectrum was extremely wasteful for voice communication. There is actually very little data transmitted, although an entire channel or circuit was tied up in the process. The answer for digital data would be to break up the transmission into small packets, each with an address. The system would then look for any available channel that had capacity between voice transmissions. It would send the individual packets by whatever channel was available and then re-associate these at the receiving end into a continuous message as originally transmitted. This process was known as packet switching. In May 1990, CANTEL launched the Mobitex System, which allowed subscribers to exchange text messages via radio waves. It was primitive but it was a start.

CANTEL Mobitex

Mobitex was an attempt to make better use of the scarce spectrum available for analogue signals. Given that analogue has the privacy problems noted earlier and the fact that it was somewhat subject to external interference (like an AM radio broadcast), the push was on within the industry to come up with a digital standard.

A complex cellular antenna

Despite the best efforts of the CTIA, this was not to be. Bell South, CANTEL and some others elected to use a digital system called Time Division Multiple Access, or TDMA. This could be thought of as the equivalent of FM broadcasting, and the digital use of channels was far more efficient than the old analogue signals. Many others in the industry chose to use Code Division Multiple Access (CDMA), which can be thought of as something of the equivalent of the Mobitex packet switching approach.

Neither would end up being the correct choice, but CANTEL did launch its digital service in February of 1992 and gradually phased out the analogue network.

Ultimately CANTEL converted its entire network to the European standard first introduced in 1991, called GSM (originally Groupe Speciale Mobile and later standing for Global System for Mobile Communications). This conversion, although expensive, put CANTEL in the lead in digital cellular in Canada.

Controlling Cantel

We left the story of ownership of CANTEL with the investment of Ameritech, who took up 19.9% of the shares at the end of 1984. In many ways, this only worsened the control situation. While Ameritech was the most understanding partner we had, as they shared the same challenges of developing a new cellular entity in the U.S., there were now four partners giving direction to the company. CANTEL had only until December 11, 1985 to find additional Canadian equity. My preference was to have one of the founders put up the Canadian money and become the dominant shareholder. At least this would help CANTEL gain some coherent direction from the top. Ted did make an offer to buy the entire $15 million private placement. As I knew there were going to be competing bids, I approached DOC to get a two-month extension until February 28, 1986. Philippe did decide to make his own bid of $15.9 million. Ted, with only four days before the February 28 deadline, raised his offer to $21.1 million and added an offer to buy shares from Belzberg and Beaubien as a "$2.1 million cash inducement for each of them to concede."

Again, I commend the reader to review the details in Chapter 14 of *High Wire Act.* It is a fascinating story. The result was that Belzberg backed Rogers' bid.

By May, 1986, First City and Télémédia became minority partners with 12% of the equity and 6% of the votes. Ameritech's stake remained unchanged.

Ted then sold most of the stock to the public company, RCI, who ended up with 38% of CANTEL's fully diluted shares.

Now that RCI was the controlling shareholder, First City and Télémédia really wanted to get out. A Share Exchange Agreement dated May 29, 1986 purported to give the minority shareholders a 'put', which would allow them to convert their ownership into Class B shares of RCI if CANTEL had not gone public with at least a $12.5 million public float. One of the stipulations was that CANTEL had to become a Reporting Issuer before December 1, 1987. This led to Ted's brilliant ploy of filing a Shelf Prospectus, which made CANTEL a Reporting Issuer even though they had issued no public securities. The ploy worked, and to celebrate the event, I distributed a few mock tombstones that resembled those for an actual issue. The tombstone noted that the securities offered hereunder were $000,000,000. The tombstones were of black marble and the humour also somewhat black.

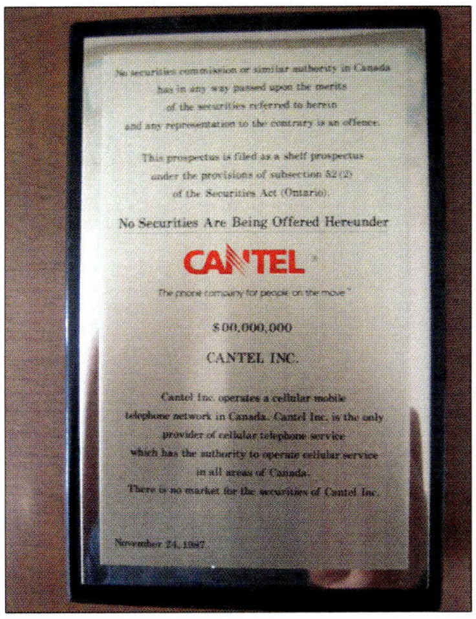

My attempt at humour with a phony tombstone

In the meantime, Ameritech was exasperated with its 'quarrelsome Canadian partners' and agreed to sell its stake in CANTEL to RCI for C$106 million. This was over four times its original investment and, again, was an example that Ted's partners, although worn down by the process, always made money. Dick Notebaret had replaced Denny Strigl as President of Ameritech and was delighted to get out of the arrangement with a solid profit.

Ted still needed to raise the $12.5 million but really had no desire to take CANTEL public. In any case, by that time the $12.5 million was not money that CANTEL needed, having negotiated a $1.1 billion revolving line of credit with the banks. The amount was so small that neither First City nor Télémédia could have sold their holdings in the market in any case without totally disrupting it.

More financing for a cash hungry company

Ted put the heat on his co-founders by hiring investment bankers Scotia McLeod, Burns Fry and RBC Dominion, to organize an IPO. I, and other members of management, embarked on a cross-country and European Road Show.

Ted then made Philippe an offer of $50/share for his stake. This was a huge gain for Philippe, estimated at $146 million. Philippe accepted.

Negotiations with the Belzbergs were more difficult but were finally concluded.

CANTEL Tombstone

On October 11, 1988, the Belzbergs signed the Agreement of Sale. We quickly called the underwriters to pull CANTEL's IPO, which was aborted just two hours before the final prospectus would have been filed with OSC.

The final ownership gyration came when RCI decided to buy out the remaining 3% of the stock held by employees, friends and some of the CANTEL owner-dealers. An independent committee of the RCI Board was formed and RBC Dominion Securities was asked to do a Fairness Opinion. They placed the value of the CANTEL stock as somewhere between $92 and $105 each, far higher than RCI's $58/share offer.

In March, 1989, Ted did raise the offer to $98.50 for each CANTEL share and the ownership of CANTEL was now 100% in the hands of RCI.

CANTEL eventually did go Public in August, 1991, when it was felt that the full value of CANTEL was not being recognized in the value put on RCI shares.

The Name Of The Game

Shortly after Ted gained effective control of CANTEL, the inevitable happened. Ted decided that to get more continuity with the other Rogers services, the company should be renamed Rogers CANTEL.

At the time of the original listing

This came as no surprise to me, as I had gone through exactly the same process with Premier

Cablesystems in Vancouver. When Rogers acquired control, the name was rapidly changed to Rogers Cable and the management was integrated to form a national cable company.

Even the name Rogers Cable had an interesting background. Prior to acquiring Premier, Ted had made an unfriendly takeover of Canadian Cablesystems. The latter was the name that was used initially for franchising in the United States. Ted wanted to change the name of all the cable operations to Rogers Cable.

The motion to do this came up at an Annual General Meeting of RCI. There was an animated discussion from the floor, as a number of shareholders doubted the wisdom of dropping the name of Canadian Cablesystems. Ted appeared somewhat embarrassed in defending the proposed change.

I intervened noting that:

• The Rogers name had a long and prestigious association with the communications business, going back to Rogers Majestic and earlier companies. Rogers was a logical name for the new organization.

• Canadian Cablesystems was not a good marketing name in the United States and the Rogers name would not have the connotation of a foreign company bidding for licences.

• Finally, I noted that the only alternative that was considered was Fierheller Cable and even I would not vote for that!

There was a ripple of laughter in the room, and with no further objections the motion passed.

There were going to be many more changes. Although the CANTEL name would continue to be used for marketing purposes, the legal name of the company had been changed to Rogers CANTEL Mobile Communications Inc.

Although it is beyond the timeframe of this book, following the joint venture established by RCI and AT&T, it was concluded that the brand name for CANTEL's products should be changed to CANTEL AT&T. This was shortly repositioned to Rogers AT&T Wireless, and the name CANTEL slipped into oblivion.

In August 1999, British Telecommunications Inc. and AT&T Corp. jointly acquired a 33.3% equity stake in Rogers AT&T Wireless. In January 2000, the new corporate identity was launched, using the name Rogers AT&T Wireless.

The final transition occurred when AT&T and BT sold their positions back to

The final trade when CANTEL went private January 6, 2005

RCI, paving the way for Ted to acquire Microcel Telecommunications Inc., a move that was not favoured by his partners. It led, however, to the wireless company becoming the largest in Canada. Microcel had over one million wireless users and had a soundly established marketing name (FIDO) for its service. When the proposal to buy Microcel was running into a lot of negativity at the Rogers Wireless Board, I spoke out strongly in favour of the move, noting that the cost to acquire that many subscribers would have been far more than the cost of the acquisition, even assuming we lost a few in the process.

The move took an aggressive competitor out of the game and enhanced the Rogers Wireless position in Canada.

But now back to the CANTEL story.

PART FOUR

Room at the Top

THE FLORIDA FLING

By October, 1988, CANTEL had exceeded 100,000 subscribers as noted. I felt it would be useful to have a strategic planning session with all of the senior management to look ahead for the next five years. We elected to have this combined with an upbeat, morale building session at the Doral Golf Resort in Florida. We even went as far as to hire a 'motivator' to organize some upbeat sessions.

The latter turned out to be a total waste of money. As usual, the group needed no inspiration. When the motivator watched the group after dinner on the first evening forming a Conga Line and chanting "the place to be in '93", he just laughed and gave up with his efforts.

The next evening ended with a water volleyball game in the pool. The game went on into the wee hours until the hotel management finally closed the party down.

We knew we had a Mission – and we were the Missionaries.

However, as I would be in my 60s by 1993, and as our forecasts indicated we would have more than 500,000 subscribers by that time, it was time to augment the leadership.

The Doral Resort – scene of 'the place to be in '93' session

NEW NAMES FOR THE 90S

As the company grew rapidly, I had expressed to Ted that I felt we needed a Chief Operating Officer. Many of the things I had led – the dealings with government, the rollout across the country and the buildup of staff, had been accomplished. The job ahead would be much more a management process.

Ted and I had lunch on the patio at the Granite Club in the spring of 1989 to discuss this. Ted proposed Jim Sward.

Jim was a respected manager as President of Rogers Broadcasting. I knew Jim well and felt he would be an excellent choice.

Jim was one of the trio of senior operators at RCI at the time. Colin Watson was President of Rogers Cable, Jim was President of Rogers Broadcasting and I was President of Rogers Wireless.

At an Annual General Meeting the prior year, Ted had been very complimentary noting that "these are the three stools on which the success of Rogers depends." I think he meant to say "the three legs on the stool." In any case Colin leaned over to me and said, "I think he just called us three turds."

Ted agreed to meet with Jim, who was somewhat reluctant. I followed this with a meeting with him at the King Edward Hotel to outline what I thought was involved in the next few years in running CANTEL on a day-to-day basis. We agreed that splitting the job between a Chairman and CEO and a President and COO would work and he accepted.

At this point it is worth exploring a bit the relationship that the three of us had with Ted.

In the numerous management meetings that the four of us held, Ted seemed to have ongoing confrontations with Jim, who was a bright and strong-minded individual.

Colin and Ted also had a few confrontations.

I seemed to be reasonably immune to this. Perhaps it was because Ted's background was in cable and broadcasting, and therefore he tended to let me run CANTEL as I saw fit. Or perhaps it was just that we had a long history and he seemed to appreciate what I

Jim Sward, President & COO with me, 1990

had done to give him his big boost in the cable business by supporting his acquisition of Premier. Whatever the reason, although Jim did a sound job of managing CANTEL, he did not even make it until 1993. Although Jim was only 47, he and Ted agreed it would be best if he retired, which he did in March 1993.

By 1993, CANTEL had exceeded its original objective of 500,000 subscribers with a base of 573,000.

Jim left the Rogers organization entirely, being hired by Izzy Asper as CanWest Global's President and CEO.

The search for a new President began but this time with a difference. As we were almost certain to go outside the organization to look for a new President, Ted and I both felt that he or she should be President and CEO, and I should retire from active involvement in the management of the company. Ted would take over as Chair of Rogers Wireless and I would move to being Honourary Chair of the Wireless Board. Ted also graciously appointed me one of the two Vice-Chairs of Rogers Communications Inc., the other being Ted's long-time associate Phil Lind.

Ted with David Gergacz

There now began a game of musical chairs. The search turned up an American, David Gergacz. David had an impressive résumé (that Ted circulated widely) including being a jet fighter pilot during his military service. He may have had some of the Walter Steel syndrome. David was a good marketer. His major contribution to CANTEL was launching a brand within a brand called AMIGO.

An Amigo phone display

The AMIGO concept was simple. As phones were now becoming largely hand-held, it made sense to find a new way of packaging the product. The AMIGO phone was sold off-the-rack with a pre-authorized connection and the whole product was presented in a blister pack. It truly moved the product to a consumer item. A special pricing package was developed and the customer could walk out of the store with a working phone. The concept was very successful.

However, as had happened with Walter, David and Ted did not see eye to eye on a number of matters, and by 1995 he had been replaced by Stan Kabala.

Stan was, in turn, replaced by Charlie Hoffman by 1997.

By that year the company's revenue had reached

$1,241 million, with an operating profit of $386 million. We now had more than 1.8 million customers.

Charlie brought in a number of American managers and despite the revolving door at the top, the company continued to develop rapidly.

Charlie, who was 50 at the time, had a problem. His wife did not want to relocate to Canada and he ended up flying home every other weekend. He decided to leave. This left Ted with the challenge of finding yet another President.

The choice was one of the happiest for the organization.

The decision was to hire Nadir Mohamed, who was then the President of Telus Mobility. Nadir was only 44 at the time.

Stan Kabala 1995

Charlie assisted in the recruitment, but it was Ted and Nadir who hit if off right away.

Nadir has subsequently led Rogers Wireless through one of its most productive periods, giving it continuity at the top that it had not had for some years.

Nadir subsequently moved up to replace Ted as President and CEO of Rogers Communications Inc. when Ted passed away.

Charlie Hoffman, 1997

Nadir Mohamed

Rogers Wireless Today

I suppose it is not surprising that a company that launched into a new field with such high growth potential would go through many management, ownership and other changes. Despite all of these, Rogers Wireless has come out on top.

It now represents 55% of the total RCI revenue of $11.6 billion.

With 8.2 million subscribers, it is Canada's largest wireless company, covering 94% of the population.

It has been a tumultuous but rewarding 25 years.

It was fun being part of the process.

Another Opening of Another Show

At the Ericcson facility with Nick Kauser, John McLennan and David Parkes

'Family' get-together with Nick and Judy Kauser at the back left,
Kathy Parkes at the front left and Glenna and me at the back right. David Parkes took the photo

PART FIVE

Back to the Beginning

The Ted Factor

At the beginning of this story, I said I would try to come up with some thoughts on why a start-up company with no background whatsoever in the telephone business could take on some of the largest and most profitable companies in Canada, and ultimately become the largest factor in the wireless telephony field in Canada.

It is tempting to give a self-serving answer. I could, for example, point out that we had:

• attracted a young, devil-may-care staff that showed amazing loyalty over the years
• exploited some wonderfully innovative engineering through Nick and his group
• developed an imaginative set of product offerings and some innovative pricing
• did not hesitate to use aggressive marketing.

As an example of the latter, at a very early stage we had an independent engineering firm do a comparison of our service with Bell. The results indicated that, at least at that time, we were offering a better service with fewer dropped calls. We quickly followed this by taking out ads headed "Clearer Than A Bell."

CANTEL at 10

Probably one reason we were successful is that, like AVIS, we were number two and tried harder.

In a way, it was easier for us because no one expected that we could take on the Telcos, and therefore we had nothing to lose (except the founding shareholders' equity!).

But let's be realistic.

The real difference was Ted.

The title of his book, *Relentless*, tells it all.

Ted was driven to succeed and he had the ability to make it happen.

In addition to being bright and relentless, he had the reputation of being fair. Even after he had outmaneuvered his partners, they expressed admiration for his imagination and drive.

He knew from the start that he wanted 100% control of what became Rogers Wireless. He would work day and night to stay ahead of his partners to get the control he wanted. I described the Shelf Prospectus as a way of outmaneuvering the PUT. I could just as easily have described how he researched that the only jurisdiction in Canada that would allow a Canadian company with American directors to be construed as foreign investment was Nova Scotia. Ted reincorporated what was then CANTEL as a Nova Scotia company and appointed some American directors to prevent the Belzberg's or anyone else from taking up Ameritech's position in the U.S.

While trying at times, it was an experience working with Ted that could not easily be duplicated. He was favoured with a prodigious memory, incredible work habits, an innovative imagination and he tied all this together with his well-known drive to succeed.

Time With Ted

My relationship with Ted was also somewhat unique. We were friends but were not really close socially.

I was, of course, involved in all the appropriate Rogers events, many hosted by Loretta at their home on Frybrook. But largely we travelled in different circles – his was a world of business contacts and he seldom wasted time on those who he felt were not contributing toward his business objectives. My social contacts were more those from the not-for-profit sector and various industry communities.

Still, the relationship worked.

At the Rogers' Hearing for Premier: from the left George, John Graham, Ted, Phil Lind

Ted was almost oblivious to time. To the chagrin of my wife, Glenna, he would for example call on a Sunday night at say, 11.00 p.m. to discuss something he was working on. The fact that he got us out of bed simply never occurred to him.

I remember one occasion when he called and said he was really interested in some ideas I had put in a memo. I could not even remember the memo but when he described it I said, "Ted, that was nearly a year ago!" "It is still a good idea," replied Ted, and there was no point in objecting.

But the really important calls from Ted were infrequent but very personal. They occurred when he just wanted to talk to someone he felt he could relate to.

I would get a call at home, often on a Sunday, and remember one from Lyford Cay, when he was having his worst health problems. He wanted to outline his plans for setting up an Office of the President with Nadir, Alan Horn and Bill Lynton, just in case.

On an earlier occasion, he wanted to ask my advice on how best to introduce his son Edward to the business. He felt Edward should start at the bottom to really gain an understanding of cable in particular. I agreed, but suggested that Ted use his contacts in the United States with other cable companies to find an appropriate position where Edward could learn the ropes, without being put in the position of having to learn from those he would subsequently manage. Ted accepted this advice and I am not sure that Edward ever knew that he had even asked me.

I am sure Ted did the same with others whose opinions he respected. Ted made very good use of his Board. It may seem strange that someone who had full voting control of a company would still go to great lengths to ensure his Board was with him.

True, he had some members of the Board who were 'friends and family', but he had some outstanding external Directors such as John Tory, Sr., Robin Korthals, Gar Emerson, Peter Godsoe, David Peterson and many others I could mention.

Ted and George in front of the Temple of the Gold Pavilion, Kyoto

Some of Ted's Team:
Phil Lind, Ted, Gar Emerson and Graham Savage

Ted loved the debates at Board Meetings, and I believe he provoked them from time to time. While he would rarely admit that he was wrong, if he felt the counter-arguments were valid, he tended to go along with them.

Another huge reason for the success of RCI, and hence wireless, was Ted's ability to command an unbelievable loyalty from his management team.

As noted earlier, there was no shortage of disagreement and, on many occasions, the members of the management group were left seething with frustration.

But this only indicated that Ted had the ability to attract and retain a group of independent thinkers to whom he once again really listened.

I could not begin to outline them all, but Phil Lind, Alan Horn, Graham Savage and the late Bob Francis come to mind.

There was a sense of camaraderie amongst the group and they were not above joking about the style of the company. When Ted made a large donation to Ryerson University for a new Business School, the management group dubbed it the Rogers School of Micro Management.

I should add that, as Ted pointed out in *Relentless*, his relationship with his step-father, John Graham, was unique. John was also something of a mentor to me, and certainly an example of how a true gentleman can make his mark in business.

But Boards, management and partners notwithstanding, it was Ted's personal drive that led to the development of one of the most successful communications companies in Canadian history.

On a personal note, I miss him greatly.

Ted Rogers and Family: Melinda, Lisa, Ted, Loretta, Martha and Edward

APPENDICES

Appendix A

OFF-BASE OFF-SHORE

When I finally stepped down as Chair and CEO of what is now Rogers Wireless in 1993 on the arrival of David Gergacz, I moved into the head office operation as one of the two Vice Chairs, as noted earlier. Ted and I each viewed this as a transitional phase to retirement.

There were, however, a number of matters external to RCI management that Ted asked if I would either continue with or take on. Ted simply did not have the time to do many of these things, but felt that RCI needed an 'ambassador' to help improve the profile of the organization, amongst other things. For example, together with his stepsister, Ann Graham, I was responsible for the charitable donations for Rogers.

Ted had been one of the founding members of an organization called SMART Toronto but really did not have time to run it, although he was the Founding Chair. I took this on for him.

I was also the Chairman of the Information Technology Association of Canada, although I stepped down in June of 1994 (on next page). I also gave innumerable talks when Ted just did not have the time.

I was also taking over the position as Chair of the Board of United Way of Greater Toronto on May 31, 1994.

There was one set of activities, however, that deserves a bit more attention.

Swimming In A Bigger Pond

Ted had always viewed himself as being a potential major player on the international scene. He had the experience of the very successful expansion of cable franchises in the United States, led by Phil Lind. In fact, it was the ultimate sale of these U.S. assets for $1.1 billion that enabled him to finance the expansion of wireless in Canada.

CANTEL had developed a very good reputation in North America for providing a reliable network, even though the management was never satisfied that it was good enough (or perhaps we had a good network because we were never satisfied with it). We also had one of the longest cellular networks in the world.

Ted felt we could lever this with some offshore ventures.

The Velvet Revolution

The first opportunity arose in 1989 with The Velvet Revolution in Czechoslovakia. The new Czech government realized that if they were to be a success in the commercial world – after two generations of occupation, first by the Nazis and then by the Russians – they would have to start by upgrading their telephone system.

Letter from Ted re ITAC, page 1

Edward S. Rogers
President and
Chief Executive Officer

Rogers Communications Inc.
Scotia Plaza, Suite 6400
40 King Street West, Box 1007
Toronto, Ontario M5H 3Y2

Tel. (416) 864-2101
Fax (416) 864-2333

May 30, 1994

Mr. George A. Fierheller
Chairman
Information Technology
  Association of Canada
Suite 401
2800 Skymark Avenue
Mississauga, Ontario
L4W 5A6

Dear George:

Over the past number of weeks I have been receiving an increasing number of telephone calls from members of ITAC expressing increasing levels of distress on your imminent retirement as Chairman of the Association and expressing their concern on how the void that you will leave may be filled.

Fortunately, I have been able to effectively calm these concerns by pointing them to the precedent of what has happened at Rogers Communications Inc. since you "retired" as Chairman of Cantel. You have continued to provide the same informed guidance as you always did and in fact have taken on the leadership of as many projects since your retirement as you had before. These are, of course the same characteristics you displayed when you "retired" as Chairman of United Way. So from that perspective members of ITAC should have little fear of you "disappearing into the sunset".

I think the only person who has seen less of you since "retirement" than before is Glenna!

It is important to the Association that you do stay involved. George, you have brought so much to the Information Technology Industry in Canada. From your early days with IBM and STL to your move to Premier Cablesystems you have always been an innovator and leader. As the founding President and Chief Executive Officer of Cantel Inc. you have truly earned the title of "Father of Cellular" in Canada.

Throughout all these roles you have displayed a facility for bringing people often with competing and diverging views and interest together. It has been this strength above all others I think that has made your career such a success.

George, I look forward to your continuing contributions to ITAC for many years to come and congratulate you on the achievements and progress that you have achieved at the Association during your term as Chairman. You are truly a great Canadian and a great, great friend.

Kind regards.

Yours sincerely,

Ted

Edward S. Rogers

/kh

Letter from Ted re ITAC, page 2

The existing system was all old Soviet technology with mechanical switches. During one of my first visits to Czechoslovakia, I was given a tour of their Prague facility. It was the first mechanical relay-based switch I had ever seen and the clacking of the relays, as they connected caller to recipient, was awesome. The installation had not been upgraded since the 1950s. It was immaculately maintained but could serve only about 10% of the population.

The solution was to move directly to wireless.

Within weeks of the establishment of the new government, under Václav Havel, it was decided to put out a tender for a state-of-the-art cellular system.

CANTEL had developed a good working relationship with McCaw Communications, and Ted and I saw Craig McCaw fairly regularly.

Ted and I had also established a contact with Ian Vance, Managing Director of British Telecom. CANTEL already had considerable business dealings with Nokia, as they were one of our telephone suppliers. Ericsson was the logical choice to provide the network if we decided to put together a consortium.

Electromechanical Telephone Switch

These companies did form a partnership to reply to the bid. Bob Berner, one of Nick Kauser's senior engineering team, and I took the lead for CANTEL.

Photo taken on Wenceslas Square

The team arrived in Prague only a few weeks after the coup. The government offices were in a small building at the foot of Wenceslas Square – it was a four-storey building with no elevator.

The 'ministers' were all new and, as I recall, the Minister of Communications had been a school teacher a few weeks earlier.

The office equipment consisted of a few old manual typewriters.

The first meeting was a revelation about a revolution. Our consortium was amongst the first to make contact with the new government and one of their first requests was that we help them debug their buildings. The Communists, who were still very active on the political scene, had placed bugs in all office buildings, hotels and, of course, on the phone system itself. We arranged for some debugging equipment to be sent to them from a supplier in New York.

The next request was even more bizarre. They felt that Canada was a wonderful example of a democracy in action and asked if we could give them advice on how best to run an election. I asked how many parties might be running and they replied "it could be 20 or 30." I suggested that a large chalkboard might be the best way to start rather than trying an automated approach, but I did put them in touch with Canada's Chief Electoral Officer for more professional advice!

Language was a problem. The then current generation had been brought up speaking only Czech or Russian. There was very little English spoken. Fortunately, we retained the services of a very bright young girl to be our interpreter.

We also had a consultant from Canada whose job it was to coordinate the team. This turned out to be a bit of a disaster.

At the first official meeting of the bidders with the Ministry of Communications, our 'consultant' started off by asking what tax concessions we could expect. I listened for a short while and then interrupted by asking what they really needed most, and how we could become a partner with them to meet these needs.

I suggested, for example, that we could assist with language training, as essentially all manuals were in English, and could also use the facilities of some very fine educational institutions in Canada to conduct management training courses.

The whole tenor of the meeting turned around at this point and suddenly we were talking as friends, not just as participants in the bidding process.

Our young interpreter had likely embellished some of the things I said and afterwards she laughingly noted that she knew exactly where I was going and why.

Celebrating the Velvet Revolution

It was going to be a fierce competition with bid groups from around the world and we needed, in my opinion, to differentiate ourselves by being truly responsive to the needs of a struggling new government.

We made good use of our Consular contacts. The British Consulate was helpful in providing all kinds of statistics but really did not get their hands dirty in the process. The U.S. Ambassador, at the time, Shirley Temple Black, and the Americans held a reception for us. It was the Canadian Consulate, however, who really worked with us. They set up meetings, attended 7:00 a.m. breakfasts and everyone remarked they were truly helpful.

In the end, we were not the winner.

One of the Baby Bells got the business with a multi-million dollar grant package that our group was unwilling to consider. However, we learned a lot from this effort.

• First it is difficult to get decisions with such a large and diverse group (this reminded me of the early days of CANTEL)

• Not all, including CANTEL, were really behind the bid. Most of the team members paid lip service to the bid process, but it really came down to Bob and me to try to get the job done.

• We learned that consultants cannot do the job for us.

However, a new opportunity was about to present itself.

Duo In Deutschland

We were to try one more European excursion. A bid was being called for a new cellular system in southern Germany.

Bob and I concluded that we would be better to go with a nearly all-German group. One of the partners was Shell in Deutschland, based in Hamburg. Another was the large German conglomerate Salzgitter, and to this we added a local agricultural co-op serving the area to be licenced.

But we still made the mistake of using a consultant as the coordinator. He was a stubborn, arrogant German named Fritz who had done some work for Shell. (It is alright for us to refer to some Germans as being stubborn and opinionated as both Bob Berner and I have German backgrounds. Bob speaks German quite fluently.)

Bob Berner, Senior Vice President and Chief Technology Officer

Hamburg and the Inner Alster

Fritz would simply not listen to our exhortations to use aggressive forecasts. As a result, our financial projections looked weak, relative to some other bidders. He would rant at meetings, each of which we came to refer to as a Fritzkrieg.

For a second time, we lost.

There is always some benefit to these projects. I became quite close to the Managing Director of Shell in Hamburg. He was a delightful individual and really seemed to appreciate the efforts we put into trying to win the licence. I gathered he had been wounded in the War, as he walked with quite a limp. He and his wife took me to a concert in the Church of St. Michael in which their son was singing. We all then went to a dinner at the Landhaus Scherrer. I felt rather sorry for the son, who likely would have preferred to be with the other choristers.

Even Farther Afield

At one time Ted, Graham Savage and I toured a couple of countries in the Far East with the aim of both looking for investment and possible partners. Ted's reputation got us into meetings with Li Ka Shing, who controlled Hutchison Whampoa based in Hong Kong, and other fascinating individuals such as Dr. Kazuo Inamori, the founder of Kyocera. Although Kyocera stood for Kyoto Ceramics and was based in Kyoto, they had extensive interests in wireless through an organization now known as KDDI. Dr. Inamori, amongst other things, had established the Kyoto Prize, a prestigious international award.

It was only a survey trip, high on interesting experiences but low on results.

Graham Savage, George, our interpreter, Ted and our hosts from Kyocera

Ted's hit and run style of doing business in the Far East did not sit well with those used to a more leisurely approach based on developing personal relationships first. I recall with amusement one meeting in Japan where Ted opened the conversation asking how the Japanese cellular company handled their accounts receivable. The reaction around the table was one of shock!

Japanese Hospitality

As those who have done some business in the Far East would know, a typical business meeting in Japan was very formal with many Japanese representatives at the table. It was customary to have interpreters and the meetings therefore progressed slowly. This did not work well for the ever impatient Ted.

At one of the inevitable social events that followed such meetings, geishas constantly refilled our saké cups and beer glasses. As everyone got increasingly loaded, our Japanese hosts started to speak in fluent English, even cracking jokes. Having worked through interpreters all day, we were left wondering what side comments we might have made thinking they would not understand.

We did use the trip to build relationships with a number of Japanese cellular phone manufacturers, who were becoming increasingly important. One of these was Mitsubishi; our contact was a senior executive known as Super Yamaguchi. He was quite a character and spoke fluent English. After a successful meeting, he invited us to his private club. It turned out he was an accomplished pianist. We had just arrived in Japan and were somewhat jet-lagged. The drinking and singing went on for hours until he finally had his chauffeur take him home. Stamina was one requirement for dealing with the Japanese.

There was one positive result from all this, at least in the short run. During an earlier trip to Australia to deliver a paper, I spent some time with Alex Arena, who was with the Australian Telecommunications Authority. He had just been appointed Director General of Telecommunications in Hong Kong. He encouraged us to enter the paging business in Hong Kong. We did this in partnership with a Taiwanese company called Pacific Telecommunications Limited. The launch was quite successful and we quickly obtained 12,000 subscribers.

However, while we provided some technical advice, it was really difficult to get a management handle on what was being done in Hong Kong from our office in Toronto. It was also hard to understand the financial reporting and we had the uneasy feeling that the group was into some trading operations in the PRC that might be taking more management attention than we had expected. It was not unusual for Taiwanese companies to operate in the People's Republic of China through Hong Kong. Ultimately we sold the operation.

View from the Chiang Kai-Shek Memorial Hall – Taipei

Ted's Rogers Wireless was now once more essentially an all-Canadian operation.

The Grand Tour

However, Ted does not give up easily. I was asked to give a paper at the Pan-Asian Mobile Communications Summit in Singapore in July 1993. Ted asked if I would do a more comprehensive assessment of what, if anything, Rogers should be doing offshore.

Rather than repeating what is already in a document, I have attached my trip report, which concluded that RCI should not waste money trying to develop offshore interests. I concluded that:

• The RCI Board and most of the RCI management had no stomach for such ventures, which involved a reasonable degree of risk with people with whom they were not comfortable.

• We were not willing to really be there, and unless we chose to make a significant local investment and apply the resources to make it a success, we were better to stay away.

• It was largely an expensive ego trip.

Instead, I suggested that RTL, Ted's private company, invest as a founding partner in a new venture being set up by Charles Sirois called Télésystem International Wireless Corporation (TIWC).

The old and new in Kuala Lumpur

Télésystem International Wireless Corporation

TIWC had a mandate to develop wireless systems in countries around the world. It had a dedicated staff headed by Bruno Ducharme (whose father had been on the RCI Board). They had a Board ready to devote time to the project and they had what appeared to be adequate financing from the Caisse and others.

Ted took my advice and made the investment. He joined the Board himself. I ended up being quite involved and Chaired their Audit Committee as well as an Independent Committee when the time came for some financial restructuring.

The story of TIWC should be written as it was a fascinating example of a good idea that led to rapid over-expansion followed by the financial backing falling apart and the whole project collapsing.

TIWC rapidly developed systems in places as diverse as Hunan Province in the PRC, Brazil, Mexico, India and several countries in eastern Europe.

The latter turned out to be good systems but most of the rest were major headaches.

RTL finally sold its shares and fortunately made money on them.

With Some Chinese Government Officials, New Territories, Hong Kong

Bruno was a very

competent and hardworking executive. He and I still keep in touch as we both serve on the Board of the Canadian Institute For Advanced Research.

My contact with Charles Sirois expanded even further as I served on the Board of Teleglobe, as a representative of O.M.E.R.S., for a good part of its hectic history in the 1990s.

In summary, going offshore requires that everyone in the company be on-side. To be successful, offshore ventures require the same dedication and intensity as CANTEL had during its start-up within Canada.

It is either that or forget it.

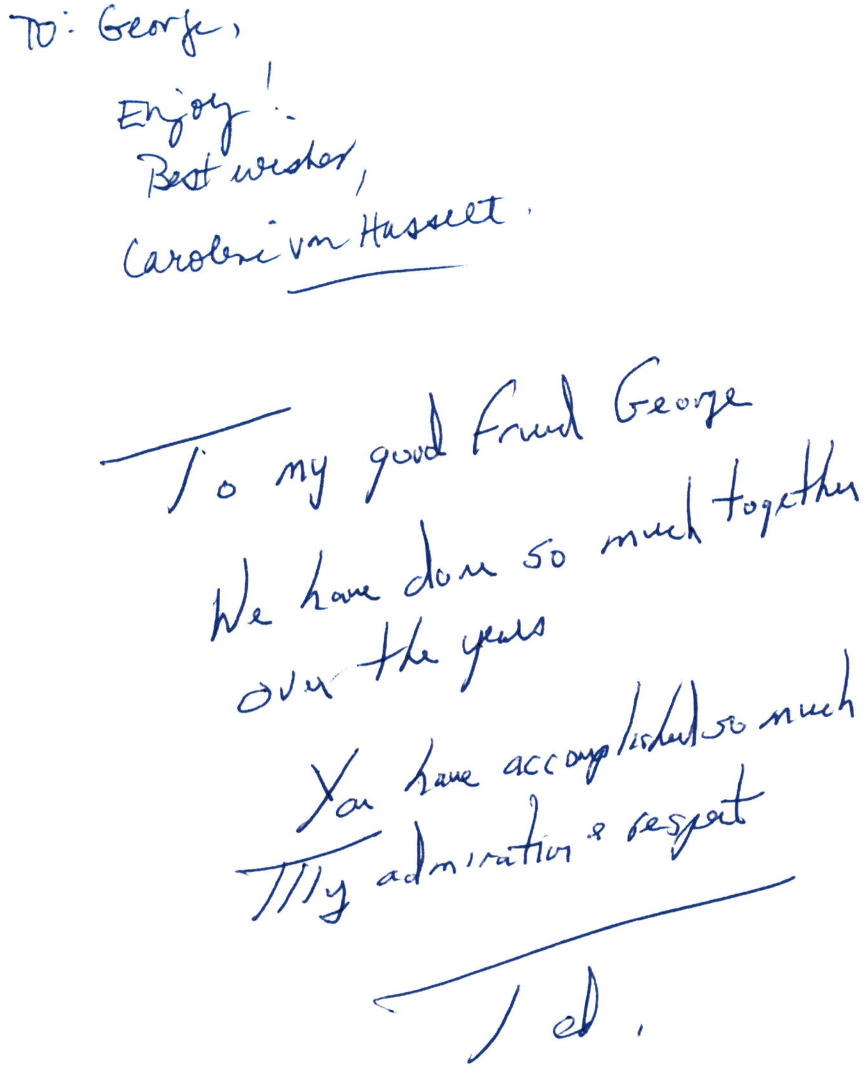

To: George,

Enjoy!

Best wishes,

Caroline von Hasselt.

To my good friend George

We have done so much together over the years

You have accomplished so much

My admiration & respect

Ted.

The Dedication page in my High Wire Act book written by Caroline and Ted.

Appendix B

PROSPECTS IN THE PACIFIC RIM

July 1993

George Fierheller

REPORT ON PROSPECTS IN THE PACIFIC RIM

George A. Fierheller

July 1993

The attached report details the outlook for specific countries, reports on our current investments and notes the contacts made.

Overview

The potential for mobile communications is everything we have been led to expect. However, everyone else has recognized the same potential. There is little virgin territory.

The exceptions are the People's Republic of China (PRC), Taiwan, and to a lesser degree, Thailand and Indonesia. Malaysia looks like a very attractive country in which to do business, but the field is already crowded. Vietnam is a wild card.

RCI Options

Assuming we want to be involved in this potential, we have the following options:

1) Status Quo. We can stay with what we have in the Northern Pacific Rim area. DDI requires no assistance and is a portfolio investment only. PTL, on the other hand, does need some technical assistance under this or any option.

2) Expand on our own. This will take a meaningful investment in people, as well as dollars.

   In addition to the technical and marketing expertise we would need to provide, we would also need a development group on the ground in the area, e.g. Colin Russel, Rick Siemens.

   We would need to expand our current contacts with organizations such as Star Paging, Pacific Concord, Hong Leong, the George Ho Group.

   We would likely also want to be involved with one or more of the large non-Asian operators in the area.

3) Augment what we have with portfolio investments.

   These would be the types of investments we could make through Charles Sirois or Millicom.

   3.1 If we make the investment just for cash, then we need put up no people.

   3.2 If we claim expert advice as what we put up for equity, with or without cash we will have to deliver on both.

4) Drop PTL We could let our investment in PTL be watered down or simply sell it and go with either 3.1 or 3.2.

5) Drop PTL and stay only with the DDI portfolio investment. This would be effectively withdrawing from the area.

Comment on the Options

1) If we stick with PTL, we should support them with enthusiasm and provide some limited technical support to protect our investment, e.g. 3 - 4 man-months per year.

It will not help our reputation in the area or with PTL, if we slow them down by dragging our feet (we would start to look like CP at Unitel).

If we provide the support, this is a credible option.

2) Going on our own is a non-trivial effort. We would need a proper team, such as those we would put in place to get any Canadian licence. To this, we would have to add ongoing operational support. This would likely involve considerable senior RCI management time.

I question if RCI is really large enough to do this and win against the huge firms already in the area, e.g. Singapore Telecom, or the international players already heavily involved, e.g. Cable and Wireless, British Telecom, the RBOC's.

I would only recommend this if we go with the McCaw/AT&T Group and really let them run the operation and we take a small percentage.

3) If we buy in to a Millicom portfolio, we are late in the game and would not be getting 'founder's stock'. The first round of real wealth creation would be already gone.

A better approach would be to work with Charles Sirois, who already has people in these areas, e.g. Malaysia. There is still some room for a Canada Inc. approach. Canada is highly regarded in the area, and the Canadian Government is well equipped to help here.

More important, we bring expertise that Charles does not have.

I like this approach best.

4) There is likely no reason to drop PTL, even if we go with 3). In fact, having people from Teleglobe/Telesystem and our people in the area would only enhance our investment in PTL

5) Withdrawing from the area with this much potential seems drastic. We have already made the effort to be there.

   Also as Albert Gnat pointed out during his visit, PTL has done well in establishing their paging operation and are being very innovative in work in the PRC. I would suggest that even the Status Quo is better than total withdrawal.

Recommendation and Action

I would recommend that we go with 3).

We need to put some dollar estimate on what we are willing to invest in the area.

The next step would be an early internal meeting, which should include Albert as well as you, Graham, and me.

This could then be followed by a meeting with Charles Sirois.

SOUTH-EAST ASIA

DOING BUSINESS IN ASEAN

Overview

The countries that are part of the Association of South-East Asian Nations have a population of over 350 million people.

South-East Asia is quite different in business terms from the Northern Pacific Rim countries, e.g. Japan, Taiwan, Hong Kong, Korea and the PRC. Business in the North centres around Hong Kong. In the South, the centre is Singapore.

There is little trust between the two groups and great rivalry.

RCI has its two interests in the North. What follows is a report on South-East Asia. The principal countries are Singapore, Malaysia, Indonesia, Thailand, The Philippines, and Brunei.

► REPUBLIC OF SINGAPORE

Singapore has a population of 2.7 million in an area of 639 sq. km. It sounds like an ideal market for cable T.V. or mobile communications. In fact from RCI's standpoint, it is neither.

It is a very tightly run economy and Singapore Telecom has a locked-up monopoly until 1997. They already have a fully loaded AMPS System with over 60,000 subscribers, an ETACS System with over 50,000 and are about to launch a GSM System.

Their cellular prices are so low (US $0.12 per minute) that even if they did open up competition after 1997, which is unlikely, there would be no market.

They have launched a very successful CT2 operation where the per minute costs are half the above. I met with Wong Hung Khim, the President and CEO of Singapore Telecom, and a number of his senior staff in a private meeting to discuss possible joint ventures outside Singapore. Mr. Wong, although a friendly individual, politely told me that we had nothing to offer and they were already in joint ventures in a dozen countries. They clearly see themselves as the dominant force in the area.

Their Chairman, Koh Boon Hwee, is one of the young tigers of the area. I did not have an opportunity to meet him, but I gather he is Harvard educated, about 36 years old and a ball of fire.

The only bright spot was Charles Cosgrove, Group Director, Business Development. Charles is from California and in charge of all overseas joint ventures. I spoke to him privately after the meeting and he indicated that we should not be quite so discouraged. He actually from time to time runs into overseas operations where a Canadian involvement could be helpful. It is not our expertise that they are looking for however, but rather Export Development Corporation, loans or guarantees. There is no equivalent to this in Singapore.

However, I would not hold my breath. I believe that there is nothing to be done with the telephone company.

Government Linked Companies (GLC's)

Obviously, Singapore might be used as a base for joint ventures in South-East Asia. For this purpose, there are two types of companies operating in Singapore. The first are quasi-government organizations not unlike crown corporations, but with public shareholdings. The Government controls these through an organization called Temsek Holdings. This company also makes investment offshore.

I did not have time to directly contact 2 or 3 of the main companies. These are by and large widely diversified organizations and might easily be interested in joint ventures, although I suspect Singapore Telecom will have them tied up. Principal amongst these are:

Singapore Technologies Industrial Corporation

This is headed by Mr. K.S. Wong, who is a McGill grad. He could be a very interesting contact.

Keppel Corporation

This is also a diversified group, but definitely into telecommunications. Interestingly enough, their Chairman is Philip Lam, who is a Canadian.

Sembawang Shipyard

This sounds like an odd possibility, but they are again into many high tech projects. There are no known Canadians or contacts in this organization!

This lead me then to the second group of companies that are simply large, private or public organizations. The only one I had time to contact was the:

Hong Leong Group

I met with John Wong, Group Managing Director. John was a former head of IBM in Singapore.

Hong Leong is a huge private corporation, although many of their subsidiaries are publicly traded. They are widely into real estate, have the largest financial house in Singapore, and such miscellaneous investments as a major string of Hotels in Asia including the Grand Hyatt in Taipei.

Peter Chi of Peat Marwick accompanied me on the call. John brought in K.T. Wong (no relation), General Manager of HL Technologies. It was soon hard to tell who was selling what to whom. I indicated our interest in possible joint ventures in South-East Asia in mobile communications. John immediately suggested that their best contacts were in Indonesia and they were fairly shortly going to be bidding on a major paging operation in Jakarta. K.T. Wong was somewhat less enthusiastic about this possibility, as he was more interested in selling us portable cellular phones.

One of their subsidiaries manufactures phones and they have an AMPS compatible unit that sells for under U.S. $300 for a small hand-held unit. I am getting some details on this.

Hong Leong is a solid contact, and certainly John Wong realizes some of their limitations in what they are trying to do. They are selectively approaching various markets and believe that Indonesia represents a major potential. They will forward some details to me on this project.

► MALAYSIA

On Saturday, July 10th, I flew to Kuala Lumpur (yes, the industrious Malaysians work on Saturdays!). There I contacted Mr. Tremain Tanner, who is the Canadian investment counsellor with the Malaysian Industrial Development Authority. This is a group that is funded by CIDA. Their main purpose is to arrange for partnerships in Malaysia on behalf of Canadian companies.

Malaysia has about 18 million people, English is widely spoken, it has essentially an English legal system and impressed me in being a top spot should we be looking for an investment opportunity in the area.

The only problem is, we are hardly the first in line.

Telekom Malaysia is privatized and the Government regulates the area through the Department of Telecommunications. They have a well established NMT-450 Cellular System with over 90,000 subscribers. A competitive system was licenced to a company called Celcom in 1989, and they installed a 900 Megahertz ETACS System that now has over 115,000 subscribers.

A third cellular carrier is expected to be licenced in 1996, likely on the GSM Standard.

As expected however, our friends at Hutchison and others will be lining up for this. It will be a very competitive bidding process.

There are no particular foreign ownership restrictions, although it would be impractical not to involve the Bumiputras (the Malay Malaysians).

At the Pan-Asian Mobile Communications Summit, I had a discussion with Dr. Dzaharutin Mansor, Manager of R&D for Celcom. He is a most impressive individual and generally reinforced my view of Malaysia as being a real comer in South-East Asia.

► INDONESIA

With a population of 188 million, Indonesia should be the sleeping giant of the Asia Pacific area. Unfortunately, it appears to be largely still asleep.

The population is spread over some 13,700 islands. It boasts the second oldest cellular operation in the region, and yet has only 0.01% penetration.

The reason is the abysmally high cost. There is an upfront installation cost of $500. The monthly cost of $20 and the minute price of $.20 is not out of line, but the terminals are very expensive, e.g. $5,000 for a portable. This makes the cellular system the most expensive in the world in terms of start up cost.

As a result, cellular has only been sold to oil company executives and government officials.

To be more specific, there are four companies offering cellular. Even by the admission of the Director of Engineering for PT Telekom, the quality offered by all of them is poor.

These companies offer three systems at the moment:

- NMT 450 with 16,000 subscribers and saturated
- AMPS 800 with 23,000 subscribers with some growth potential
- TACS 900 with 25,000 subscribers, which is a dead system

► BRUNEI

This is a tiny city-state in the northern part of Borneo. The population is only 250,000. It is a sultanate.

It is also a very rich area with the wealth being based in oil.

It has a primitive telephone system. Although, I have never been there, it is described as having homes on stilts, and yet the inhabitants have two cars, vcr's, satellite dishes and want and can afford the latest in technology.

There is a 'one stop agency' called the Ministry of Industry and Primary Resources, which would need to approve any telecommunications activities in the country.

The drawback is that the only real way to do business in the country is to become friendly with the family of the Sultan. This may not be as difficult as it sounds, but to say the least, this is a long shot.

► VIETNAM

Vietnam is referred to as the 'Wild East'. It has a population of about 80 million, and could well turn out to be a very interesting marketplace. In fact, it is relatively politically stable.

However, the GDP/Capita is very low and there is little affordable demand.

The French are very active in the area with Alcatel being the dominant player.

Interestingly enough, Sinclair Roche Temperley, lead by Franca Ciambella, has a contract funded by CIDA to implement a Canadian-like court system. This should have the effect of getting some kind of commercial law into the area.

The people are industrious, but the country is starting from a long way back.

Not likely a starter for us.

► AUSTRALIA

While not part of the current examination, a brief update may be useful.

Ian Slattery, General Manager, Competition Policy for AUSTEL, gave an interesting presentation.

What used to be Telecom Australia is now TELSTRA. This followed their merger with the Overseas Telecommunications Corporation (OTC).

Australia has elected to go GSM. There are now 2 licenced competitors. The first is Optus, which is a competitive telephone company, as well as competitive mobile supplier. They have been reselling the AMPS System and now have about 15% market share.

Vodafone will start in Australia as the third cellular licencee in late 1993. All three will be offering GSM.

Interestingly enough, Australia intends to totally phase out the analogue system within a decade. This is despite some 700,000 customers now in the country.

As I forecasted after my trip to Australia a couple of years ago, the cost of entry is unbelievable. Vodafone paid about $140 million for its GSM Licence - glad we stayed home!

• GENERAL CONTACTS

Singapore

I met with Gavin Stewart, High Commissioner for Canada, and his staff. Gavin is recently arrived (8 months), but is certainly well connected and was most helpful. The Commercial Counsellor, Jim Feir, was very knowledgeable and together with Frank Quah, the Senior Officer, set up the appointment for me with Singapore Telecom.

Frank's wife is the secretary to one of the Senior Directors at Singapore Broadcasting Corporation (Lam Lychow, Director, Engineering). A brief discussion with him however indicated that there is little opportunity for cable in Singapore. There is very tight control on content, and therefore little need for multiple channels. They are worse than the CRTC!

Auditors

Thanks to Bill MacKinnon, I met with Bobby Chin, Managing Partner of KPMG Peat Warwick, and Ng Boon Yew, a partner specializing in the telecommunications area. They were very helpful in setting up the meeting with Hong Leong.

Legal Firm

The prize contact for the whole South-East Asian area however is Franca Ciambeila. She is a Canadian lawyer, from Montreal with the firm of Sinclair Roche & Temperley. They are one of the largest legal firms operating in the area. Beyond that however, she is the Chairman of the Canadian/Singapore Business Association. I had breakfast with her (too bad it was not dinner - she is gorgeous!). She is a mine of information and would be really helpful in setting up whatever further contacts we may need.

In the past, we have used Baker Mackenzie, who are good, but are only half the size of Sinclair Roche in Singapore. Baker Mackenzie also happen to be the lawyers for Pacific Cable and Wire in Taiwan.

ITU

I met with Gary Brooks, who was formerly with DOC, but is now stationed in Geneva. He is a member of the Radio Regulations Board of the International Telecommunications Union. He is a good contact on the Asian scene, at least from the regulatory stand point. He confirmed the impression I am already getting that Asia is essentially going GSM (except for Korea, which has gone CDMA). It does not look as though our experience with TDMA will count for a lot in South-East Asia.

Cheryl Tritt

I spent some time with Cheryl, who was formerly the Chief of Common Carrier Bureau, Federal Communications Commission. She is now partner with Morrison & Foerster in Washington. She referenced a paper put out by the United States, which cautioned the industry in having two digital standards. Apparently, they are very concerned about the impact on export markets if the U.S. itself cannot get its act together between CDMA and TDMA.

Cheryl could be a good contact for us, if we are doing regulatory work in the U.S.

She also noted that, in her opinion, it would be early 1996 before PCS licences would be granted in the United States.

She was further optimistic that mobile data would be a $3-4 billion industry by the end of this decade.

Kwok Shu Wong

I had dinner with Kwok, who is the Assistant Director (working for Alex Arena), of the newly established Office of the Telecommunications Authority (OFTA) that has just been set up July 1st, 1993 in Hong Kong to replace the situation where telecommunications was regulated by the Post Master General. Kwok confirmed the intention to licence yet another cellular network in Hong Kong. As he said, there are five networks now, and all are profitable so why not (he slightly overstated the case, as obviously Smartone with their GSM System are only getting launched).

He confirmed that there would be no restriction on foreign owners, no requirement for local manufacturing and no need for local partners (although, our experience has indicated that this is most desirable).

They are also going to provide the cellular operators with direct access to the international gateway (HKTI).

NORTHERN PACIFIC RIM

► HONG KONG

I had dinner with my friend, Alex Arena, the newly appointed Director General of Telecommunications for the newly established Office of the Telecommunications Authority (OFTA). Alex had met with Wellen Sham and others of PTL and participated in the opening of their operations. He shared my high regard for what they have accomplished in the short space of time.

In general, Alex's approach is to open up the Hong Kong market to as much competition as is reasonable. However, it is not obvious that there is going to be a fifth cellular licence. Alex instead advised that he is going to put out a Request For Comments later this year asking for advice on what the next move should be in Hong Kong. It is possible that they may wish to licence some advanced form of PCS (Alex was quite interested in the CT2+ Canadian operation), instead of just assuming that more cellular is the answer. Obviously, we should reply to this.

I probed Alex on the likelihood of a successful PTL bid, given that they are not Hong Kong based. He quite honestly replied that this would have no effect on the technical evaluation, but given the 1997 turnover to the PRC, involvement in the PRC would be the most important factor given reasonably equal technical bids. He noted that Warf had won the right to cable television in Hong Kong, even though the licence period extended beyond 1997 on the basis that that organization had been heavily involved in the PRC in real estate and other ventures.

No doubt the involvement by the George Ho Group in real estate and other ventures in the PRC helped the Smartcom bid for the GSM licence in Hong Kong. He noted that the two China situation might be a negative, but he believed that the gradual assimilation of Taiwan would lead to a three China approach, e.g. the PRC, Hong Kong and Taiwan following the 'one nation, three systems' approach. As long as the Taiwanese were well involved in the PRC, there should be no problem. Fortunately, PTL's involvement in paging in China and their increasingly large involvement in other projects should put them in good stead for a Hong Kong bid. I was advised by Philip Wong that Pacific Capital, for example, is involved in a U.S. $4.2 billion real estate development project in Fujian. He felt that PTL were doing all the right things to be a major player in both the PRC and Hong Kong.

Wellen told me that it was their intention to involve the MPT from the PRC in their Hong Kong bid. This would of course be immensely helpful as well.

As an aside, Alex dumped all over involvement in countries such as Indonesia. He said that the corruption and need for political pay-offs had discouraged even the most adventurous Australian companies from doing business in the area.

PACIFIC TELECOMMUNICATIONS LIMITED

In Hong Kong, I met with Philip Wong, Director of Pacific Capital (Development) Limited, Wellen Sham, Francis Cheung and others of the PTL Group. I was most impressed with what they have done in a short time to set up a paging operation in Hong Kong.

There are now over 2,000 subscribers and should be over 12,000 by year end. They believe the 12,000 will be a break-even point.

They forecast being at 30,000 subscribers, which will essentially be capacity by the end of 1994.

Their approach is quite interesting. As Albert Gnat noted, they are taking a very conservative approach to the accounts receivable. Although, they will allow anyone to return a pager at anytime and discontinue the service, they require a HKD 900 deposit up front. They essentially provide a free Chinese-character pager, which comes from OI, a Japanese manufacturer. However in reality, the rental for the pager is built into the monthly fee of HKD 260. Customers can change models, e.g. from numeric to alphanumeric by putting up the difference in the deposit.

It is important to understand that there is a Paging Association in Hong Kong, which is simply a cartel. The cartel sets a minimum of HKD 260 per month, but allows service providers to offer extra services for that price. In Hong Kong, most of the paging companies charge HKD 330. The established companies are reluctant to reduce this amount, allowing PTL to rapidly gain market share.

The PTL pagers provide only 13 Chinese-character messages. The competition will offer more or less unlimited messages (which takes more transmit time). For most applications however, the PTL offering is quite adequate.

The two stores are clearly better than the competition (although not necessarily better than those of Hutchison). They have an aggressive T.V., radio and print media campaign with a very imaginative robot logo to promote their product together with their call number 1308.

They provide many extra services, e.g. a stock quote service, which will allow the amateur investor to select a small number of stocks and the pager will beep if their stock exceeds limits up or down.

They sell with about 20 direct sales representatives and 10 more in the retail stores. The retail stores account for about 45% of their sales with another 55% coming from the direct sales force. The sales forces are on a very small salary with the rest being commission.

They train their operators to meet the Hong Kong demands for a very personalized service. The training period lasts a week, where the competition usually trains for 1-2 days.

I was very impressed with the professionalism of the operation and have no reason to assume that they will not meet their targets.

Trading Operation

PTL has got a licence to export pagers and cellular telephones to the PRC. This has the potential of being an immensely profitable operation. It is this for which they require the HKD 30 million line of credit. Essentially, they can buy the OI pagers from Japan for about HKD 1,400 and resell them in the PRC for HKD 3,500 - 4,000. This will be a major source of PTL's income during 1994.

People's Republic of China

It is important to understand how they are getting paging licences in China. As we know, the central government, through the Ministry of Posts and Telecommunications (MPT), is trying to exert authority in this area and is coming into some conflict with the regional governments. For the Tianjin operation, which will start in August, they cut a deal with the Petroleum General Corporation, the PRC Government-owned oil company who have a private paging frequency. There is nothing to stop such an organization selling outside users on their paging frequency.

In Tianjin, with a population of over 9 million people, there are only 50,000 pagers in total. They believe they can sell 8,000 by year end. In fact, selling is hardly the word for it. People line up to get paging, given the shortage of frequencies.

As Tianjin is essentially an estuary area, it is very flat and the implementation of a public network using the private frequency is quite inexpensive. This should be an extremely profitable operation in the very near term.

Cellular in the PRC

This is another odd situation. Cellular licences are available for AMPS Systems as these frequencies are basically unoccupied. To the extent that there was any standard system in the PRC, this has been a TACS-900 System. However, there is still some confusion about who can get licences from whom.

Apparently in Chengdu, PTL had got the approval for a cellular licence, but in an adjacent area someone also got approval and highly publicized this to drive up the stock price in one of the new PRC public companies. This infuriated the regulators and they temporarily stopped all granting of AMPS licences.

The mayor of Chengdu apparently still will get them the cellular licence.

They have several other projects pending in the PRC. Evidently, these are sufficiently profitable that they will get a one year payback of capital.

Action

I am very impressed with what our partners have done. We came very close to blowing our relationship by our insistence on written confirmation of their oral agreement to include us in all future ventures. I was assured by the group and had this subsequently confirmed by the High Commissioners Office for Canada that oral agreements have the force of law in Hong Kong.

On this assumption, I have approved the expanded line of credit and reconfirmed the injection of equity.

I still believe it is essential that we provide some on-the-spot interface for this group, as we can be of significant assistance. They will share the cost of people provided, and I believe this can be an immensely profitable operation in the very short term.

I will investigate whether Watson Zan could be our best way of assisting them in the forthcoming PCN licencing in Hong Kong, amongst other things.

STAR DIGITEL LIMITED

Star Digitel is a wholly-owned subsidiary of Star Paging (International Holding) Limited. Star is the second largest paging company in Hong Kong, having been in operation for over a decade. They are very financially successful and are considered a good quality supplier.

One of our senior technical people at Cantel, Watson Zan, is a good friend of Terence Tarn, Director and General Manager of Star Digitel. During a recent trip to Hong Kong, they got together and Tarn indicated they would like to talk to Cantel about some possible future partnership. I met with Tarn while in Hong Kong. He was well aware of our relationship with PTL, and incidentally, said that he thought PTL was doing a first class job in launching paging.

He indicated he had been in touch with Wellen Sham some months ago suggesting, possibly in the PRC and elsewhere, the two companies should form a joint venture. He indicated that PTL had shown no interest, feeling that the two companies were essentially parallel, both had good contacts in the PRC and there was not much to be gained by putting the organizations together. However clearly, Star Digitel feels the need of a good paging operator (who am I to disillusion them!) as well as someone who has cellular experience, particularly with AMPS 800 and digital.

I might add that Star Digitel has about 20% market share in Hong Kong. They also indicated that doing exactly what PTL has just started to do in the PRC, i.e. getting the use of existing paging frequencies. They now have about 20 cities available. This is hard to verify, but certainly they have been very active.

I raised this with Wellen and got the same response. They are just not anxious to share the equity and believe that there is enough potential for both companies operating in parallel.

It does not appear that this will get very far, but I was quite impressed with Terence Tarn. They would certainly be an alternative to PTL, not that I am recommending this.

▶ TAIWAN

Taiwan has a population of about 21 million. They introduced cellular in 1989 through the Directorate General of Telecommunications (DGT) making this one of the youngest cellular networks in the Asia/Pacific region. The initial system is an AMPS 800 Standard. It essentially covers only the three major metropolitan areas of Taipei, Kaohsiung at the south of the island, and Taichung in the middle.

At the moment, they have about 390,000 cellular subscribers for a 1.9% penetration, which is very high considering the youth of the system and the generally poor service.

The DGT has done little in the way of privatization or opening the market to competition. There is no indication when this will happen.

PACIFIC ELECTRIC WIRE AND CABLE

I was met by Tapin Hu, Specialist in the Planning Division of their Head Office Group, who worked on the initial cellular application for Hong Kong. He took me to their Ta-chi plant where I met with Chris Pon-Yean Lee, Vice President, Engineering and Manufacturing.

It is an impressive plant that manufactures all kinds of power and telecommunications cables and has recently set-up what looks to be quite an advanced fibre optic plant. They export essentially to South-East Asia.

From everything I could gather, they are highly regarded in Taipei as one of the leading industries. They are quite profitable. In 1991, they earned NT $13 billion with a profit of NT $709 million.

The more important meeting took place over dinner with Jack Sun, President, Wellen Sham, Tapin Hu, Chueh-Pin Chang, a consultant who used to work for the DGT and is now helping them with telecommunications proposals, and finally Philip Wong, one of their Directors who I had met in Hong Kong.

It was a very good meeting, as we had solved the financial problems in Hong Kong. I explained to Jack that our Board was concerned about the lack of communications between the two companies and the lack of adequate financial reporting. This had lead in turn to our request for more definite commitments in writing from them. However, we understood and accepted that they really could not commit on anything in Taiwan as there is no way of knowing what rules will be set for foreign ownership (see below) or even when cellular might open up in the country. They felt that this was an unfair request, and in retrospect, I agree with them.

I reviewed my discussions with Star Paging, Pacific Concord, Alex Arena and anything else I felt was useful. I wanted the exchange to be very free and open.

It was agreed that until the Taiwan situation had clarified, we would continue aggressively in the PRC and Hong Kong. Their feeling was that we should not dilute our efforts beyond this as these were the major markets. I explained that I had been on a tour of South-East Asia, and it was possible that we might look at other joint ventures in those areas, if Pacific Electric Wire was not interested. They felt this was fair.

They did specifically ask for some assistance for a few weeks in the PCN area. They believe that PCN licences could be available in the PRC, where frankly they would make a lot more sense than they do in Canada. Here everyone is spoiled by good cellular systems, but in the PRC with limited coverage, limited distance systems might be very acceptable, bearing in mind that there are still 40 major cities with no telephone system at all.

If you agree, we likely could provide them with that assistance, which would be paid for by PTL.

As an aside, Pacific Electric is likely going to be an investor in Iridium. This system really seems to be taking-off in the Pacific Rim countries.

In any case, after many toasts, I believe the relationship is on a good basis once again.

CANADIAN TRADE OFFICE IN TAIPEI

The next day I met with Ronald Berlet, the Director of our Trade Mission, and Andrew Chiang, their Commercial Officer. They were reasonably helpful, although I did not feel that they were up to the calibre of either Hong Kong or Singapore. Bear in mind that we do not recognize Taiwan, and hence this is not a High Commission but rather just a Trade Office.

I tried to get a good outside impression of the stability of the Taiwanese Government situation. As you know, the KMT (the Kuomintang of Chiang Kai-Shek) is still in power, but the DPP Opposition Party has been pushing strongly for an Independent Taiwan, rather than gradual integration with the PRC. The reality is that there is no chance of this happening. If Taiwan ever tried to become truly independent, they would be invaded without question.

There are over 10,000 Taiwan companies operating in various ways in the PRC having put up an investment of nearly US $15 billion. This is funnelled through Hong Kong or other vehicles where the economic ties are extremely strong with the PRC.

As an aside, Taiwan is a strange country relative to many others in the area. There is essentially no British influence, and therefore they drive on the right like the rest of China, speak very little English other than at senior levels and are more oriented toward the U.S. than any other non-Chinese nation. Knives and forks are as common as chopsticks!

Ron agrees with Jack Sun's assessment. It could be a year or five years before the Government decides to open up telecommunications with the exception of cable. Literally the day I was there, a Bill was tabled to allow a private cable television operation to be established. There is already an extensive illicit system in Taipei. At the time of tabling, no foreign ownership was going to be allowed. However, the KMT is going to amend that to allow 20% ownership. If we are interested in this area, there is a major Taipei satellite and Cable conference coming up September 15-16th.

Had I more time, Ron could have arranged for me to meet with the Minister of Transport and Communications. Chao-Shiuan Liu is a graduate of Sherbrooke and got his Ph.D. in Chemistry from the University of Toronto. I believe he would be a great contact, but it could just not be arranged on such short notice. It is interesting to note that his brother is involved in a major scandal for influence peddling to get contracts for Ericsson with whom he previously had worked.

In summary then, there is really nothing we can do for the time being in Taiwan, except be patient.

▶ OTHER NORTHERN PACIFIC INTERESTS

SOUTH KOREA

South Korea has a population of about 44 million people and has had one of the most rapid growths in the cellular industry anywhere in the northern region. Cellular was introduced in 1984 and became an immediate hit with foreign firms operating in the country.

Korea Telecom has had a monopoly to date, and now has over 300,000 AMPS-800 subscribers.

In 1992, a licence was awarded to a consortium of Vodafone, Hutchison and GTE.

As you aware, South Korea recently announced its decision to go CDMA and plans one of the earliest roll-outs anywhere.

There is likely no role for us to play here.

JAPAN

We will get a more complete reading on the DDI Cellular operation from the meeting with Mr. Onodera on July 26th.

It is interesting to note, however, the progress being made by DDI. At the end of 1992, they had 363,000 subscribers vs. IDO's 310,000. Their growth rate is much faster. NTT of course still dominates the market with over a 1 million subscribers.

There are certainly no giveaways in Japan. Even the off-peak rate is about US $0.36 with the peak rates all exceeding US $0.50. There are no free minutes!

Japan is going its own way on digital. The Japanese Digital Cellular (JDC) Standard is based on TDMA. As usual, these decisions are made for political reasons, rather than technical. No one else in the Asia area will likely use JDC, and Japan will push this to try to protect its own market. The same of course applies for the North American Standard(s) and the European push for GSM.

We should be aware that the MPT has announced it will issue two competitive digital cellular licences for each of the eleven regions. Each new carrier will therefore be competing with one digital carrier, NTT's national licence and an analogue competitor.

Appendix C

THE CASINO GAMBLE

JULY 1993

GEORGE FIERHELLER

ROGERS™

Memorandum

Rogers Communications Inc.
Suite 2600
Commercial Union Tower
P.O. Box 249, T. D. Centre
Toronto, Ontario M5K 1J5

STRICTLY PRIVATE & CONFIDENTIAL

To: **E.S. Rogers** **Copy:**

From: **G.A. Fierheller**

Date: **July 20,1993**

Subject: **THE CASINO GAMBLE**

The Casino Gamble

I managed to squeeze in the Macau trip the evening before leaving for Taiwan. I am now an instant expert in casino operations!

Kwong Yiu Ling, Manager (Casino Operations), fortunately was in Hong Kong and he and I went over on the Jet Foil and spent the evening reviewing their operations. The background is as follows:

Macau, as you know, is essentially one big gambling casino. There is really no other reason for its current prosperity. However, prosperous it certainly is.

Dr. Stanley Ho is 73 years old, but very active in the Macau scene. He is no relation to George Ho incidentally.

He controls two companies, one of which is private. The private company, based in Macau, is the Sociedade de Tourismo e Diversoes de Macau (STDM). This company not only runs the series of gambling casinos, but is building a new ferry terminal in Macau and is participating in the second Macau-Taipa Bridge (an island just south of Macau), and has a one-third share interest in the new (and only) Macau International Airport.

By way of interest, Dr. Ho already owns the Heliport, the Jet Foil operation to Hong Kong with 15 boats, a High Speed Ferry (HSF) operation with 2 much larger fast ferries, the Lisboa Hotel and virtually everything to get people from Hong Kong or the PRC to the gambling casinos. This is really an integrated operation.

To show you the scope of Dr. Ho's holdings, I happened to pick up a newspaper ad on Nam Van Lakes. This is a U.S. $1.4 billion project to completely redevelop Macau and link it to Zhuhai.

.../2

Page 2 - Memo to ESR, re: Casino Gamble
July 20, 1993

Beyond all of this, there is a public company known as Shun Tak, which owns the Shun Tak Centre which is at the Hong Kong end of the Jet Foil/Ferry operation. Beyond this, the public company is into department stores and other real estate operations.

In a word, Dr. Stanley Ho could be considered very credible.

The Casino Operation

There are 7 casinos in Macau, and Ho is involved in nearly all of them. As there is no gambling in either the PRC (at least not legally) or Hong Kong, there are some 5.9 million visitors who come to Macau annually and they are there primarily to gamble.

As you would expect, the casino operations are run very professionally, although, I do not pretend to be any authority.

The casino scene is tightly regulated. As best I could observe and in response to my direct questions, there does not appear to be any illicit other activities surrounding the casinos. There was no sign of soliciting prostitutes, drugs, or alcohol abuse and the whole operation appeared very orderly.

There are bars, and by and large as in most casinos, the drinks are given away, but drinks are not allowed at the tables (this is exactly as is proposed for Ontario).

The worst problem they have is due to poor police protection. There have been three robberies at the tables in the last few months. I cannot see this as being a problem in Ontario.

The operation is of course extremely profitable, despite the fact the Government takes 31% of the gross.

Unlike the proposed Ontario operation, they do operate 24 hours a day. There are of course peak times, but the Chinese are such dedicated gamblers that they do go all night.

The stakes are extraordinarily high, at least by my standards. There are general tables and at most of these the minimum bet is HKD 100. The chips range up to HKD 100,000. It is interesting to note that these chips are interchangeable at all casinos (that is why one could have a robbery at one table and spread the chips around at other casinos). In Macau, the chips are just like cash in the stores.

.../3

Page 3 - Memo to ESR, re: Casino Gamble
July 20, 1993

There is also a VIP section with substantially higher stakes.

I might add that there is also a casino for the local population with lower stakes. This is in a different location, and evidently is also well attended.

The monitoring is extremely interesting and very sophisticated. As you would expect, all tables are monitored by TV to ensure there is no cheating by the croupiers. I was given a tour of their monitoring centre. It looks like the Rogers Video Centre with about 40 TV screens and a staff that has the ability to zero-in on anything that looks suspicious. They have a staff of about 30 working 3 shifts of 10 just to monitor the tables. I felt this was very expensive relative to the risk, but they assured me that if this was not done, the rip-offs would be unbelievable.

They record everything and keep the tapes for 7 days and are now experimenting with directional sound microphones. They developed their own software package together with a British firm.

In terms of the games themselves, they have many of the usual, e.g. slot machines. Baccarat is extremely popular. However, there is great interest in a series of Chinese games, including Fan Tan.

The Ontario Scene

I asked why they were interested in Ontario. Macau, of course, reverts to the PRC in December of 1999. They had a large gambling operation in the Philippines, and several years ago, this was taken over with no compensation. They are obviously worried what might happen in a few years time. They are therefore very serious about moving offshore.

I am not too familiar with the Ontario proposal, but have a few observations that I passed on to them.

Given the large Chinese population in Toronto, I suggested they should retain some of the Chinese games. These might make a very interesting addition to the casino operation.

I pointed out that people in Ontario are used to going to Las Vegas or Atlantic City, and I felt the casino operation would have to have much more glitz than the operations in Macau, which were adequate, but certainly not up to Vegas standards.

Page 4 - Memo to ESR, re: Casino Gamble
July 20, 1993

I also felt that the Ontario operation might need to be split into several tiers. They will likely be limited to only one location, but I felt that there might be a shirt-sleeve operation catering to the Chinese and other elements, a general gambling area for the middle class, and a very VIP section that might almost be a black-tie. However, I pointed out that these were only observations and a market study was certainly needed. He indicated that Richard Ling was already undertaking this.

I also noted that in Toronto most people would drive to the Casino, and therefore parking and easy access would be essential (obviously in Macau everybody walks). On reflection, it occurred to me that something like the Harbour Castle might be a good location. Possibly, one could also build a Casino on the parking lot next to the Harbour Castle (where John's Seafood Restaurant is located). There could be covered access between the two.

I also cautioned that in Ontario my understanding was that the Government would not only take a percentage of the gross (I believe 20%), but also all of the profit, allowing the operator only a management fee and expenses. Yiu said that they were aware of this and still felt this could be a very profitable operation. However, they are awaiting projections.

Recommendation

I felt the Ho Group clearly understood the business very well. Their monitoring and operational control seemed to me to be very good. However, they run a somewhat down-market operation from what I believe would be required in Toronto.

From everything I could tell, I believe it is a 'mob-free' operation, and I have no reason to doubt the integrity of the Ho Group.

I believe they would make credible partners and would have quite an attraction to the very large Chinese community in Toronto.

Beyond that, I can only say that I talked the United Way out of doing a joint venture in a casino operation (I have no problem with their using funds gained in this way, but just felt that it was totally the wrong image for them to project). I have to say that I question whether RCI should be involved. Given the current crime problems in the Chinese community in Toronto, although this may be more Vietnamese, I believe there could be some problems down stream.

We should be very careful not to risk broadcast licences or our reputation by getting involved in a venture like this.

George G. A. Fierheller

Appendix D

THE AUTHOR'S ACTIVITIES

Mr. Fierheller graduated from Trinity College at the University of Toronto with an Honours Degree in Political Science and Economics in 1955. He joined IBM in Toronto that year and subsequently progressed through a number of positions in their sales organization. He was Marketing Manager for IBM's federal government business in Ottawa, prior to founding Systems Dimensions Limited (SDL) in 1968.

Mr. Fierheller was President of SDL from the inception of the company until it was acquired by Crown Life of Toronto. SDL was one of the pioneering companies in the computer services industry in Canada.

In April 1979, Mr. Fierheller moved to Vancouver as President and Chief Executive Officer of Premier Cablesystems Limited. In July 1980, Premier merged with Rogers Cablesystems Inc. to form one of the world's largest cable TV companies. Mr. Fierheller was a Vice Chairman of Rogers Cablesystems Inc. and Chairman of Canadian Cablesystems Limited, as well as the President and CEO of Rogers Cable TV – British Columbia Limited.

During 1983, Mr. Fierheller led the team that was successful in winning the mobile cellular radio licences for CANTEL. He was the founding President and CEO of CANTEL Inc. In September 1989, he was promoted to Chairman and CEO of Rogers CANTEL Mobile Inc. He was Vice Chairman, Rogers Communications Inc. until 1996 and now heads a private investment and consulting firm.

Mr. Fierheller has been actively involved in community affairs in Ottawa, Vancouver and Toronto including: Chairman of the Board of Governors of Carleton University; Chairman of the Finance Committee of the Board of Governors of Simon Fraser University; Chairman of United Way Campaigns in Ottawa in 1972, Vancouver in 1981 and in Toronto in 1991; President of the Canadian Information Processing Society; member of the Executive Committee of the National Arts Centre; a Trustee of the Vancouver General Hospital Foundation; a Director of Vancouver Opera; and a member of the Vancouver Centennial Commission.

Since returning to Toronto, Mr. Fierheller has served as Chair of the Board of the United Way of Greater Toronto; Chair, Information Technology Association of Canada; Chair, SMART Toronto; Trustee of the McMichael Canadian Art Collection; President of The

Toronto Board of Trade; Director, Ontario Exports Inc.; Chair of The Spirit of Leadership Campaign, Trinity College, University of Toronto; Chair of the Sigma Chi Canadian Foundation and President of the National Club; and was on the Campaign Cabinet of the Canadian Opera House Corporation.

Mr. Fierheller currently serves as the Chairman of the Honourary Board of the Greater Toronto Marketing Alliance. He is currently Chair Emeritus of the Sunnybrook Health Sciences Centre Capital Campaign; a Director of The Canadian Institute for Advanced Research; Past Chair, Toronto Adventurers Club; and is on the Board of Business for the Arts and SOS Children's Villages in Ottawa.

He has also served on a number of public company Boards including: Extendicare Inc., Falconbridge Inc., Telesystem International Wireless Inc., GBC North American Growth Fund, and Rogers Wireless Inc.

Mr. Fierheller has received many awards including a Doctor of Laws degree from Concordia University in 1976, a Doctor of Sacred Letters from Trinity College in 1999, the Award of Excellence from the Canadian Wireless Telecommunications Association; is a Significant Sig from Sigma Chi Fraternity and a Member of the Order of Constantine. In March 1991, he received Toronto's highest honour, The Award of Merit. In 1998 he received the highest award from the United Way of Canada, the André Mailhot Award. In the Fall of 1998, he was admitted into the Canadian Information Productivity Hall of Fame, as well as receiving the Arbor Award from the University of Toronto.

In July 2000, Mr. Fierheller was appointed a Member of The Order of Canada.

In May 2001, he was honoured by the Association of Fundraising Professionals as the Outstanding Volunteer of the Year during The International Year of the Volunteer. In 2002, he received the Queen's Golden Jubilee Medal and the Salute to the City Award for service to the City of Toronto. In 2005, he was made a member of the Sigma Chi Hall of Fame.

In 2008, he received the Family Service Community Award and the inaugural Sunnybrook Award for Volunteer Service.

Mr. Fierheller is the author of several books including *Finnie's Family*; *Let Me Say This About That; Do Not Fold Spindle Or Mutilate; I Gave At The Office;* and *Talk of Toronto.*

George at Trinity College

Appendix E

Bibliography

IF YOU WANT MORE!

More About Ted

High Wire Act: Ted Rogers and the Empire that Debt Built
Caroline Van Hasselt
John Wiley & Sons, 2007

Ted Rogers: Relentless
Autobiography with Robert Brehl
Harper Collins, 2008

Ted Rogers
Bruce McDougall
Burghar Books, 1995

A Bit About Bell

The History of the Telephone
Herbert N. Casson, 1910
Re-published by Titis Digital Publishing, 2008

Long Distance Please, The Story of the TransCanada Telephone System
Edmond B. Ogle
Collins, 1979

Person to Person, The International Impact of the Telephone
Peter Young
Granata Editions, 1991

The Deal of the Century, The Breakup of AT&T
Steve Coll
Atheneum, 1986

Pa Bell: A. Jean de Grandpre & The Meteoric Rise of Bell Canada Enterprises
Lawrence Surtees
Random House Canada, 1992

Wire Wars – The Canadian Fight for Competition in Telecommunications
Lawrence Surtees
Prentice Hall Canada, 1994

The Really Old Days

The Making of Mankind
Richard E. Leakey
Michael Joseph Limited, 1981

Chauvet Cave, The Art of Earliest Times
Jean Clottes
University of Utah Press, 2003

Mysteries of Ancient Ukraine: The Remarkable Trypilian Culture 5400-2700 BC
Krzysztof Cuik
Royal Ontario Museum, 2008

History of Western Music
Burkholder, Grout & Paliska
Folio Society, 2008

A is for OX, A Short History of the Alphabet
Lynn Davies
Folio Society, 2006

Communications Technology

Future Developments in Telecomunications
James Martin
Prentice Hall, Second Edition, 1977

Wireless Telephony
R.A. Fessenden
Smithsonian, 1908
Reprinted by Lindsay Publications

Cellular Communications Explained, From Basics to 3G
Ian Poole
Elsevier Ltd., 2006

The Victorian Internet, The Remarkable Story of the Telegraph
Tom Standage
Walker and Company, 1998

Wireless Telephone and Wireless Telegraph
Charles Ashley and Charles Hayward
American School of Correspondence, 1912